Great British Weather Disasters

Great British Weather Disasters

Philip Eden

continuum

Continuum UK
The Tower Building
11 York Road
London SE1 7NX

Continuum US
80 Maiden Lane
Suite 704
New York, NY 10038

www.continuumbooks.com

First published 2008

British Library Cataloguing-in-Publication Data
A catalogue record for this book is available from the British Library.

ISBN 978-0-8264-7621-0

Designed and typeset by Kenneth Burnley, Wirral, Cheshire
Printed and bound by MPG Books, Cornwall

Contents

Foreword

On the evening of Monday 23 July 2007 I found myself, against my better judgement, sitting in the green room next door to the *Newsnight* studio, waiting to deliver some pearls of wisdom about the awful floods which had engulfed towns and cities along the River Severn during the preceding 48 hours. I had, coincidentally, just completed my first version of Chapter 6 on summer floods, and over that exceptionally wet weekend I was getting that inevitable sinking feeling that the chapter would now need extensive revision. One thing that had already struck me, during the June flooding in Yorkshire, was that nobody was attempting to place these events into sensible statistical and historical contexts – in fact it almost seemed as if there was no one left in the country who knew how to do that. It was all 'unprecedented' and all 'the result of global warming' and it was 'a wake-up call to the effects of climate change'. I knew, then, that 'context' was what I had to try to tell the *Newsnight* audience about.

In the green room with me was, among others, the junior Environment Minister, Phil Woolas, whose job was to explain what the government was going to do about it. He was practising his answers to likely questions in front of the rest of us, and that word 'unprecedented' kept cropping up over and over again. So I wagged my finger at him and told him off, saying: 'If you use that word once during your interview, I will give chapter and verse on at least a dozen precedents when I go in to do mine, and that will make you look at best poorly briefed, at worst plain foolish.' What happened? Well, his first answer in the studio began: 'I'm not saying these floods are unprecedented, but . . .' One small victory, I thought. Sadly, it was no more than that. Woolas's colleagues, notably Hazel

Blears, the new Communities Secretary (Housing and Local Government, it was called, when I was learning about politics), Hilary Benn, the Environment Secretary, and Gordon Brown, the Prime Minister himself, all uttered the 'U' word during the subsequent fortnight. Doing so is, of course, the soft option for politicians, because it allows them to divert responsibility for what they can claim to be an unforeseen catastrophe.

So if my narrative occasionally veers off into a polemic against absence of context and government spin, you will understand why. And this, too, explains why Part 2, the twentieth-century chronicle of severe weather events, is so important, because it will allow all my readers to find that context for themselves. And if any of my readers are politicians and journalists, so much the better.

PHILIP EDEN
Whipsnade, April 2008

Part 1

Chapter 1

Setting the scene

How much do people remember of past weather events? Not a lot, it seems. What does stick in the mind falls into two groups: catastrophic or record-breaking individual days, and outstanding seasons. But asked to identify the most memorable weather of their lives, a small and unrepresentative sample of friends and relatives came up with only a handful of examples. Many chose 'Michael Fish's hurricane', otherwise known as the Great October Storm of 1987; those who were old enough recalled the drought and hot summer of 1976 and the big freeze of 1962–63; and there were scattered mentions of 'the day we had the wrong kind of snow' (February 1991), 'that really hot day we broke the record a few years ago' (9 August 2003), while just one spoke about the Boscastle flood (16 August 2004), but that was only because her Auntie Joyce had a friend who had to be rescued, and she had heard all about it – several times. An elderly relative did much better than all the others, recalling vividly the cold and snowy winter of 1947, the great London smog of December 1952, and 'a really scary thunderstorm in the late 1950s' (5 September 1958). Needless to say, other professional meteorologists and weather enthusiasts were avoided because many of them – especially the latter – remember every thunderstorm and snowfall that ever happens.

It was interesting to note that most of this group remembered half-a-dozen or so events during the previous year to eighteen months, including the hot summer of 2006 and the associated hosepipe bans, the London tornado of December 2006 and the Birmingham tornado of July 2005, the four-day fog just before Christmas 2006, and the big black cloud which followed the explosion at the Buncefield oil depot at Hemel Hempstead in December

2005. But once a weather phenomenon has reached two years old it seems to fall out of the human memory bank. Weather, therefore, has to be truly exceptional before it sticks in the mind.

This simple exercise prompted a rather larger and more scientific study. Fifty people were given a list of twelve events (six dramatic days and six record-breaking seasons) including two bogus ones, and asked to note down if they remembered them, or for younger respondents if they had heard about them. The results were as shown in Table 1.1.

Table 1.1 Responses to weather disasters

Event	Yes	No	Not sure
The Lynmouth flood of August 1952	29	17	4
The great London smog of December 1952	33	14	3
The great October storm of 1987	45	2	3
The Manchester tornado of May 1988 (bogus)	6	22	22
The Burns' Day storm of January 1990	13	20	17
The Boscastle flood of August 2004	42	5	3
The snowy winter of 1947	22	19	9
The long hot summer of 1959	19	21	10
The big freeze of 1962–63	38	6	6
The historic drought of 1976	41	5	4
The cold wet summer of 1995 (bogus)	10	22	18
The floods of autumn/winter 2000–1	24	12	14

It is no surprise that people's memories improve when prompted, but the trend towards dramatized retrospectives of past disasters on television, particularly on major anniversaries, has undoubtedly helped too, bringing to the attention of many younger folk some of the meteorological excesses of past years which they otherwise would not have known about. There were, for instance, a number of programmes about the Lynmouth disaster in 2002, including one which mischievously suggested that it had happened after cloud-seeding experiments had gone wrong, and these have been repeated since, while Boscastle was revisited by dozens of television crews a year after the event. And of course October 2007 brought hours and hours of airtime marking the twentieth anniversary of the Great Storm of October 1987; Michael Fish was very much in demand and it is a racing certainty that he aims to be around, zimmer frame and all, for the fiftieth.

It is also amusing to note that some 10–20 per cent of respondents 'remembered' disasters which did not happen; nevertheless, the bogus events do stand out clearly from the rest of the examples. The so-called Burns' Day storm when violent gales swept the whole of England and Wales with gusts locally in excess of 100mph, and when 48 people died, is the great forgotten disaster of recent decades. The 1987 storm has a very much higher recognition rating, although the area affected was less than a quarter that of the 1990 gale, while the death toll was also much smaller at 18. Could it be that the fact that the 1987 event was a forecasting disaster while the 1990 one had been well predicted is the crucial element here? After all, Mr Fish's TV clip saying that no hurricane was coming recently came top of a Channel 4 poll of worst-ever TV predictions, and it must have been replayed almost as many times on British television as the clip of Neil Armstrong setting foot on the moon.

Each generation seems to have its own hot summer and its own cold winter, and people are often heard muttering things like 'They don't make summers like they used to; now I remember a real summer, back in 1959 . . .' – and this even though the present warming trend in our climate has delivered to the UK half-a-dozen summers in the past two decades as warm as or warmer than anything else in the historical record. The summer of 1976 is still widely recalled, probably thanks to the combination of prolonged heat and disruptive drought; none of the hot summers since has quite replaced it in the communal memory, although the under-40s now talk about 1995 and especially 2003. Quite possibly 2006 will take over from these in due course as the archetypal recent hot summer. Those of a certain age still prefer to talk about 1959 or 1947 (yes, that was a hot summer as well as a cold winter), while in former times 1921, 1911 and 1868 would have been all the rage.

As for winters, 60 years ago elderly aunts would describe in hushed tones the privations of 1895, and parents would revel in the memories of weeks of skating in 1929, but these all vanished after the extraordinarily cold and snowy winter of 1947. Thereafter a sort of battle has gone on between 1947 and 1963; the former laid the country low in the rationing-ridden aftermath of the Second World War; the latter was statistically longer and colder but not as snowy, and it was coped with rather better (though still not terribly well) by a nation which had 'never had it so good' during the intervening decade and a half. In the past two decades, cold winters seem to have gone out of fashion, presumably influenced at least in part by the warming trend in the UK's climate. Some folk in their thirties and forties remember the winter of 1978–79, although it was not persistently cold, and certainly not cold enough for snow to stay on the ground throughout the entire season except in upland regions. The most recent generation has no truly cold winter in its memory bank: several since the mid-1990s have had one or two notably cold spells, but nothing that would have been out of the ordinary in the more average winters of the 1950s and 1960s. It will be interesting to see, assuming that the warming trend continues, whether these relative non-events enter the communal memory for want of anything more extreme. Perhaps by the middle of the present century any winter with a single significant snowfall will be so unusual that it is remembered for a generation.

It can be seen, then, that we quickly forget all but the most extreme, the most outrageous, in respect of both events and seasons. To an extent this was always so, but a large measure of our modern amnesia stems from our growing disconnection from our weather and climate, and that in turn reflects the sharp decline in the number of people who work outdoors and whose livelihood is intimately influenced by the weather. A striking illustration of this is provided by the statistics of deaths due to lightning in the UK. A recent study, published in the Royal Meteorological Society's journal, *Weather*, has shown that the risk of being killed by a lightning strike in any one year was 1.6 million to one in the late nineteenth century; between 1920 and 1960 it had dropped to a four million to one chance, and since 1960 it has levelled out at approximately 13 million to one. The death rate is heavily tilted towards males, with men six times as likely to be killed by lightning as women. Changes in the frequency of thunderstorms in the UK during the past 200 years have been relatively small, but there has been a marked change in human activity during this period. The study's author, Dr Derek Elsom, demonstrates that the decline in the death rate due to lightning mirrors the decline in the number of farmers and farm labourers over the same period. Moreover, fewer women than men work in the open air. In fact the statistics show that before 1960 a large proportion of lightning-related deaths in Britain was among farm workers, whereas since than a small majority has been those partaking in sporting activities such as golf, football, cricket and jogging.

Hand in hand with that growing disconnection goes an increasing desire to find a scapegoat for extreme weather events. For generations, weather disasters were simply 'Acts of God'. Indeed, until the second half of the twentieth century that formulation provided a get-out for all sorts of organizations and authorities who might otherwise have been considered responsible for planning for the inevitable disruption and damage or alleviating the consequences of, say, a torrential downpour or a severe gale: insurance companies, local councils, central government, water authorities, for instance, would all invoke this divine whimsy to get them out of a tight legal corner. The phrase 'Act of God' is still used in the legal world and may still be found in some insurance policies, and it is usually defined these days as an unforeseen severe natural event such as a hurricane or a flood, or a disaster caused by a severe natural event, such as a fire resulting from a lightning strike. In most Western legal systems, however, the foreseeable results of unforeseeable causes are specifically excluded from immunity to liability. For instance, when the Texaco oil refinery at Milford Haven exploded following several lightning strikes on 24 July 1994, the disaster was regarded as the foreseeable consequence of an Act of God and the insurers had to cough up. The oil company was subsequently fined £100,000 for having failed to put in place adequate safety procedures.

Climatologists, however, will have something to say about the use of the word 'unforeseeable', and it would be instructive to see this tested in a court of law. The climate of a given place is not only defined by the averages over a long period of years of such elements as temperature, rainfall, wind speed and thunderstorm frequency – that is the usual schoolbook definition – but also by the extremes experienced during that period. Thus the climatic statistics for Birmingham tell us that the average annual rainfall in the city centre is 750mm, but also that a fall of 25mm in one hour (sufficient to cause a serious flood in an urban area) occurs on average once every eleven years. A detailed and comprehensive climatology of a particular place will provide chapter and verse on the expected frequency of damaging winds, of costly floods and of lightning strikes; thus these phenomena, although they may be rare, could not be described as unforeseen. Civil engineering projects, after all, depend on this sort of statistical analysis in order to construct buildings, bridges, tunnels and roads that will not fail in an extreme weather event which may have a mean return period of 50, 100 or even 200 years. Climate experts would still admit that truly exceptional weather phenomena that fall outside the envelope of previous experience at a given locality could still be described as 'unforeseeable', and that these will probably occur less rarely during the present era of rapid climate change. Whether the term 'Act of God' to describe an atmospheric disturbance which can be fully and clearly defined and explained by scientists is anachronistic and inappropriate is beyond the scope of this discussion, tempting though it may be.

The earliest examples to be found in the literature of unusual weather being systematically blamed on human activity were published less than a century ago, during the course of the First World War. Long before that, though, Luke Howard – often called the father of meteorology – acknowledged in his three-volume study *The Climate of London* that urban areas had an influence on local climate, raising the temperature, diminishing the frequency of snow lying on the ground, and increasing the frequency and persistence of heavily polluted fogs. Howard's *magnum opus* was published in 1833. Indeed, the recognition that London's smoke contributed to the city's dirty fogs dates back to the twelfth century. Meteorological journals really got going around 1850, but for 60 years or so there was little or nothing to be found in them blaming mankind for extreme weather events – not the drought years of the 1850s nor the heatwave summer of 1868, not the remarkably cold year of 1879 nor the severe winters of 1881 and 1895, not the Tay Bridge disaster, not the Great West Country blizzard of 1891, nor the Ilkley Flood of July 1900.

It was during the First World War that the meteorological journals first reverberated to the noisy exchange of argument and counter-argument over the theory that prolonged and heavy gunfire in Flanders was responsible for the exceptional rains of the summer of 1917 in southern, especially south-eastern, England. Since then every technological development seems to have attracted its own gaggle of complainants. In the 1920s it was radio waves and airships, in the 1930s the growth in aviation, in the 1950s nuclear testing in the free atmosphere, in the 1960s and 1970s supersonic aircraft and space exploration. Some of these advances certainly have had a limited influence on the behaviour of the atmosphere, but there is no way that they could be fingered for a particular snowstorm, gale or summer downpour. It should be re-emphasized that we are talking about the kind of routine weather extremes which have happened at intervals throughout history, but which we inevitably forget after a couple of years – the sort of weather extremes which are part and parcel of Britain's climate and which therefore do not require any sort of human intervention as an explanation.

Since the mid-1980s things have changed somewhat. Now the scapegoats are things like global warming, the ozone hole, and El Niño, and we find journalists and politicians telling us that the first of these, especially, is responsible for every storm, flood, drought and heatwave that comes along. Ozone depletion in the stratosphere above the Arctic and Antarctic is undoubtedly the result of atmospheric pollution but it is not responsible for weather events in the lowest part of the troposphere anywhere in the world. El Niño is journalistic shorthand for the El Niño Southern Oscillation and is usually therefore referred to as ENSO in the scientific literature; it is a natural climatic phenomenon, a reversal of the normally westward-flowing equatorial current in the Pacific Ocean, which has occurred semi-regularly since the end of the last glaciation at least, with a periodicity of approximately seven years. It has an important influence on weather events around the Pacific basin but in western Europe we are about as far away from it as it is possible to get, and its effects here are small and nearly always masked by other factors. There is some indication that ENSO events may become more frequent and/or more intense under the influence of the present planet-wide warming trend. As for global warming, or anthropogenic climate change as it is often called these days, it is simply not possible or sensible to postulate a direct causal link between it and individual days of extreme or unusual weather. There is one proviso, however, and that is that a gradually changing climate provides a gradually changing background against which the endless sequence of weather occurs, thus the frequency and character of our weather extremes will change over a long period of years. But these changes will be detected first of all by climate statisticians and not by television reporters doing outside broadcasts from a flooded Carlisle or a windswept Portland Bill.

There is one important difference between the early twenty-first century and the early twentieth century, and that is our susceptibility to the adverse effects of bad weather. We may quickly touch on three aspects of this: insurance, the 'just-in-time' philosophy, and health and safety.

The insurance industry often reports on the growing level of payouts for damage caused by floods, windstorms and so on, 'even allowing for inflation', it says. Some insurance companies even make the illegitimate leap of logic and say that this demonstrates that our climate is becoming 'more unstable', whatever that may mean. It does no such thing. What it does illustrate is that we are richer, we have more things that can be damaged, our houses cost more to repair (yes, even after taking into account inflation), a greater proportion of the population is insured, and we are more inclined to claim on our policies than our forebears were 100 years ago.

Many aspects of our present-day economy are based on the 'just-in-time' philosophy. This is a simple but effective business strategy which involves reducing to a workable minimum stock and other inventories. A variety of stock-control signals are used to trigger re-ordering. The system is beneficial to many types of industry – particularly manufacturing and retailing – and to parts of central and local government, because it reduces the amount of capital tied up in stock, it limits the quantity of warehousing needed and it cuts associated costs. Overall, then, it improves the business's efficiency and increases return on investment. There are, however, some drawbacks, and the weather can intervene in at least two ways. First, the point at which stock re-ordering occurs is based on historical demand levels, but the demand for many products is strongly correlated with meteorological parameters such as temperature (soft drinks and salads, for example), snowfall (various motoring products), lightning (electrical products, computer repair services), and so on. Second, just-in-time stock control requires smaller but more frequent deliveries compared with the traditional re-ordering process, and these are most likely to be disrupted by snow, frost, floods or gales. These two types of problem may interact, as when severe cold may trigger a surge in demand for de-icing products, but retailers cannot acquire fresh stock because the transportation network is gridlocked as a result of snow and ice.

'Health and safety' is a late twentieth- and early twenty-first-century phenomenon which appears to have taken on a life of its own and has become a self-perpetuating industry. Launched initially – and in good faith – by a desire to minimize workplace accidents and to provide legal redress to the victims of such incidents, it is now driven by a fear of litigation, such that schools close at the merest suggestion of an imminent snowfall, children are forbidden from playing snowballs or from making slides on an icy playground, and local authorities are moved to cut down horse-chestnut trees in case a falling conker hurts a passer-by or damages a passing car. Health and safety regulations are now widely ridiculed, but it is often those who deride loudest the so-called 'health and safety Nazis' that are most likely to take legal action if, say, their child breaks an arm on an icy footpath. In Britain, the weather is most likely to catch the attention of health and safety officers during periods of frost and snow, but also in floods, gales, heatwaves and thunderstorms.

We can see how these three aspects of present-day society impinge on our relationship with severe weather by comparing briefly how things have changed even in my own lifetime. A severe gale sweeps the country and a tree in the street falls across our front garden, seriously damaging the car, demolishing a fence and breaking some windows: today we claim on our car and household insurance policies, consider suing the local authority and probably have to wait several months before all repairs are completed and monies received; 40 years ago we had no car, we repaired the fence, we called out a glazier to mend the windows for which we may have been able to put in an insurance claim if we had a policy, and we thankfully chopped up the tree and used it for firewood. A severe snowstorm brings down power lines, cutting the electricity supply: we trek four miles through six-inch deep snow to the nearest Homebase or B and Q store, only to find that they have sold out of rechargeable halogen spotlights and portable generators which we need to power the central heating and the cooking stove; 40 years ago we had a blazing fire courtesy of the tree that fell into our garden the week before, plenty of candles purchased from the corner shop when the weather forecast suggested that snow was on the way, and we cooked by gas. On a frosty morning the car is a write-off after losing traction on an icy road and I and my passengers suffer minor injuries, so we sue the highways authority to failing to keep the road safe; forty years ago I fell of my bike, brushed myself down, climbed on again and rode off into the sunset.

All in all, then, our attitude to adverse weather has changed much more in the last 50 years than has the weather itself. In the 1950s and 1960s we just gritted our teeth and got on with it. Nowadays the first reaction to a weather disaster is 'Who do we blame?' followed closely by 'Who can we sue?'

Chapter 2

Coping with the hazard

What makes a hazard

Atmospheric hazards such as hurricanes, lightning and snowstorms are just one kind of environmental hazard, though they may trigger or be otherwise associated with oceanographic, hydrologic, cryologic and geomorphologic hazards. Examples of these related types include coastal inundation, river flooding, avalanches and landslips. All of these we may group together as weather disasters. Essentially unrelated, though still under the 'natural hazards' umbrella, are seismic and volcanic phenomena including earthquakes, tsunamis and volcanic eruptions. A further group we may call technologic hazards which are, of course, not natural but which themselves may be set in motion or exacerbated by bad weather. These include industrial explosions and serious transport accidents.

Hazards can be grouped or classified in a number of ways, not just by their cause. In academic disaster analyses they are commonly categorized according to speed of onset or longevity; indeed, those events which arrive most rapidly often have the shortest lifespan, and vice versa, although the repercussions of a rapid-onset phenomenon may last for much longer than the phenomenon itself. We may think of an aeroplane crash, an earthquake, or – in the meteorological arena – a tornado, as examples of rapid-onset events, while soil erosion, famine or drought are examples of the slow-onset, long-lived kind. Qualitatively, we may describe the former as 'dramatic' and the latter as 'insidious', but both groups can be very damaging and also very costly, both economically and in terms of human lives.

Most kinds of extreme or unusual weather can be regarded as

potentially hazardous – even those phenomena which we choose to think of as unthreatening, such as warmth and sunshine – especially if the episode is prolonged. In the UK it is sometimes said that we have a very equable and therefore rather benign climate. Natural disasters are rare compared with many other countries in the world, and it is true that we do not experience the more extreme varieties of tornado or tropical revolving storms (hurricanes, typhoons, cyclones) such as routinely affect China, India and the USA, nor do we suffer coastal or river flooding anywhere near as severe as that which afflicts Bangladesh almost every year. Extremes of temperature in British cities are moderate when compared with those which residents of Moscow and Montreal, Delhi and Dubai have to cope with. As a species we handle routine rather well, so Muscovites, for instance, are used to severe cold in winter and cope with temperatures of –20 or –30°C accordingly, whereas a moderate cold snap in Delhi with the mercury between, say, zero and 5°C – a rare event – will result in dozens of casualties. Half-an-inch of rain causes the same sort of traffic chaos on the streets of Riyadh or Abu Dhabi as half-an-inch of snow does in London.

So it is in the UK. We generally deal effectively with what we are used to, but we struggle with circumstances with which we are not familiar. So rain is rarely a problem in the UK except on those rare occasions when it falls with great intensity, but extreme heat with daytime temperatures of, for example, 32 to 35°C will quickly take its toll – of our temper, our sleep pattern, our efficiency at work, and even of the lives of the very old and very young. For any particular place on our planet, it is the extremes of the climate of that locality that will impinge on the economic and social well-being of the community, and that is as true of those areas with a benign climate such as Lisbon or San Francisco or Cape Town as it is of places with a more demanding climate such as Buffalo (arguably the snowiest million-plus city in the world), Lahore (one of the hottest) or Manila (the greatest incidence of typhoons).

It was noted earlier that the impact of weather disasters on our lives has changed in response to changes in attitude and expectation and wealth, and although we cannot realistically blame the weather itself on central or local government or their agencies, we can perhaps identify who might be responsible for the way it affects our lives. We should also be able to learn how to assess the various risks, how to protect ourselves, our families and our property, how to mitigate the effects of a disaster, and how to adjust to any losses that we may ultimately suffer.

Responses to a hazard

Professor William Balchin, the prominent geographer, outlined the human responses to natural hazards in the UK under six headings. Before the disaster strikes come planning, prevention and warning, and after it has hit, relief, reconstruction and rehabilitation. It almost goes without saying that each of these responses has both community and individual aspects.

Around the world some countries have established a government department with responsibility for most aspects of disaster planning and relief, but the ability of these departments to act effectively is often limited by budgetary constraints. In the USA, for instance, the Federal Emergency Management Agency (FEMA) was established in 1979 by President Carter, drawing together the activities of a wide range of much smaller federal and state agencies, but much of its resources during its first ten years or so were directed towards civil defence; nevertheless FEMA took the lead in providing assistance in the aftermath of Hurricanes Hugo and Andrew in 1989 and 1992 respectively. During the 1990s most of the budget was earmarked for disaster relief, with a new emphasis on proactive measures such as planning and preparedness, but everything changed in 2001. New funding was found, but a substantial proportion of expenditure was now directed towards the threat of terrorism, and in 2003 the Agency was absorbed into the new Department of Homeland Security. When Hurricane Katrina struck the Gulf coast in the summer of 2005, FEMA and its then director, Michael Brown, were roundly condemned for their tardy and inadequate response to the catastrophe which befell New Orleans; the received wisdom was that, following its takeover by Homeland Security, FEMA had been deliberately starved of money and expertise. Brown, a personal appointee of President Bush, had had no previous emergency planning experience at all. Following the political fallout of the Katrina disaster and the involuntary resignation of Michael Brown, FEMA seems to have regained some of its former credibility.

In the UK, no such national agency has ever existed. It is true that serious emergencies such as the drought of 1976 and the heavy snowfalls of winter 1978–79 led to the temporary appointment of a minister (of below cabinet rank) to co-ordinate activities – in both instances, Mr Denis Howell – but as soon as the problem had passed, the minister returned to his previous role. Under the Conservatives between 1979 and 1997 no such ministerial 'overseer' was appointed, and major disasters such as the snowstorms of 1981–82, the great October storm of 1987, and the heavy snows of winter 1990–91, were handled in fragmentary fashion by the Home Office, the Department of Transport, and the Department of the Environment.

Things changed in July 2001 following a series of problems during the preceding twelve months, including an autumn and winter of unprecedented floods in many parts of the UK between September 2000 and April 2001, the short-lived fuel crisis of September 2000, and the severe outbreak of foot-and-mouth disease between February and June 2001. Since then, most aspects of disaster administration have come under the authority of the Cabinet Office in the form of the Civil Contingencies Secretariat (CCS). A Civil Contingencies Act passed through Parliament in 2004, and the Secretariat's interface with the general public takes the form of a website called UK Resilience which at the time of writing can be found at *www.ukresilience.info/*. Detailed information about the Secretariat's purview can be found there, although certain pages are heavily infested with political spin. However, most Americans have heard of FEMA, but it is doubtful whether, as of spring 2008, one-tenth of 1 per cent of Britons have heard of either the Civil Contingencies Secretariat or of UK Resilience. Nor is it at all obvious from either of its two names what this organization is responsible for. Thus, it is already failing twice at the most basic level of communication with the general public.

Nor is it entirely clear what the CCS would do in the event of a major weather disaster. None of the notable weather events since its inception – the killer heatwave of August 2003, the severe gales in northern Britain and the Carlisle flood in January 2005, and the damaging gale across England and Wales in January 2007 – seem to have triggered any overt activity. It did, however, help to co-ordinate the UK's response to the Asian tsunami on 26 December 2005, while it is presently exercised by the threat to the UK of avian influenza (otherwise known as bird flu). To be fair, CCS has already established a framework for the provision of early warnings, increasing awareness and knowledge of the different types of severe weather, and for mitigating the threat by improving defences. It also identifies government departments and agencies with specific responsibilities. Thus the Department of Health has an important role during extended periods of extreme heat and extreme cold when the very old are at particular risk; the Department of the Environment, Food and Rural Affairs (DEFRA) will be the major player during floods and droughts, especially through its own Environment Agency and the Scottish Environment Protection Agency (SEPA); and the Highways Agency, part of the Department of Transport, clearly has an important contribution to make when heavy snow, ice, dense fog or heavy rain cause serious disruption to traffic.

Crucially, the Meteorological Office (hereafter the Met Office) is identified as the source of all warnings – at a variety of time-scales – of severe weather events. The Met Office has taken this responsibility to heart: it has provided a focus for the role forced upon them by government after the forecasting debacle of the great storm of October 1987. (It is rarely noted that, had our national weather forecasters got it right and issued timely warnings of destructive gales overnight 15–16 October, there would have been a lot more casualties, probably many more than the eighteen deaths ultimately attributed to the storm, as people tried to keep their property safe, and everyone would have been just as impotent to prevent the widespread damage and destruction as they were in the event.) Nevertheless, they have yet to pitch the frequency and style of warnings correctly, as regular changes to both aspects and repeated launching of the service testify. The main problem is one of resources: a substantial number of dedicated forecasters and ancillary staff are required to produce adequately detailed and geographically focused warnings, and to disseminate them in timely fashion, but the number of days per year when these warnings are required is necessarily small. It is a problem which is hard to rationalize. The Met Office, perhaps understandably, tried to maximize the workload of their unit by issuing all sorts of different early warnings, flash warnings, travel alerts, and so on, but this merely served to confuse, and it became impossible to identify what was a really important warning of severe weather amid all the clutter.

One important part of the CCS, although pre-dating it by twelve years, is the Emergency Planning College. Located just north of York in premises which had before 1989 been a Civil Defence College, the Emergency Planning College provides a large number of courses on emergency management covering planning for, responding to, recovering from and mitigating the effects of various types of major disaster. Among those many courses is one which lasts three days entitled 'Managing High Impact Weather', run in conjunction with the Met Office. The introductory overview of the course tells attendees that by the end of it they will, among a total of nine objectives, be able to do the following: interpret the UK's weather and identify a situation likely to produce severe weather; outline how a forecast is produced, its limitations, and how severe weather is managed; and interpret Met Office products, especially impact forecasts, severe weather forecasts, and weather and health predictions. That, one might suggest, is a trifle immodest. There are professional forecasters who have been in the business for 30 years or more who would balk at making such claims.

But what can the individual do to minimize the effects of the various hazards on his family and his property? It is all a matter of awareness – awareness of the threat itself, of the various means of avoiding it, of mitigating its effects if it cannot be avoided, of knowing where to find warnings, of what insurance policies might be available, and so on. Here the Civil Contingencies Secretariat has already been at work, encouraging the Department of Health, the Environment Agency and the Met Office to publish information packs, leaflets and web pages. A good deal of the information presently available might be regarded by the educated layman as a statement of the blindingly obvious, but, paraphrasing H. L. Mencken, it is never a good idea to overestimate the intelligence of the masses. This attempt to educate the public is clearly a work in process, and eventually it will provide a very useful resource.

Legal requirements for treatment of roads in winter

Many people think that local authorities have always been required by law to keep roads clear of snow and ice during the winter. That has certainly not always been the case – indeed, the law was actually quite unclear until a High Court case a few years ago. Following that case, widely publicized at the time, of *Goode v East Sussex County Council*, the High Court judgment removed any legal requirement to salt roads to prevent ice forming unless that ice formed from water lying on the road caused by poor drainage or badly maintained drains. Responsibility under the Highways Act 1980 is confined to the fabric of the road. Thus, if it rains, or dew forms on a road, the Highway Authority has no responsibility to treat the roads to prevent this from freezing.

This judgment fundamentally changed the understanding of the legal responsibility and introduced the possibility of a much inferior service to that which had become custom and practice. To prevent this happening, the government added the following clause to the Railways and Transport Safety Act 2003:

Railways and Transport Safety Act 2003

Chapter 20
111 Highways: snow and ice

After section 41(1) of the Highways Act 1980 (c. 66) (duty of highway authority to maintain highway) insert:

'(1A) In particular, a highway authority are under a duty to ensure, so far as is reasonably practicable, that safe passage along a highway is not endangered by snow or ice.'

A highway in this context means any part of the highway to which the public has access, including the road, footpaths, cycle tracks, and so on. The phrase 'so far as is reasonably practicable' means that the highway authority can operate within the constraints of machinery, labour and finance to determine which highways will be treated. If a decision is taken not to treat all roads, then there has to be a well-defined and acceptable means of assessing priorities.

Chapter 3

The nature of the hazard

Rapid-onset hazards

We noted earlier that weather hazards have various speeds of onset, and various lifespans. At one end of the scale are rapid-onset hazards which may hit with full force in a matter of seconds, last no more than a few minutes, but which may leave behind a degree of chaos and confusion that might take months to put right. Because of their short life, the damage-to-time equation is very high, making these events particularly dramatic: they will inevitably make headline news. As far as the British climate is concerned, rapid-onset hazards include tornadoes, severe electrical storms, hailstorms and flash floods.

A typical lightning discharge involves an electrical potential of 1,000 million volts, and in a tiny fraction of a second it will heat up the air in the luminous channel to over 10,000°C. If you heat a given quantity of air it will expand, and if you heat it virtually instantaneously it will expand explosively. That expansion results in the emission of a steep pressure wave in all directions from the discharge channel. This pressure wave is heard as thunder.

Because of the different speeds of propagation of light and sound, thunder is always heard after the lightning is seen. If the strike is very close indeed, the thunder will be heard as a single short explosive bang, just a fraction of a second after the flash. Any subsequent rumble from more distant parts of the discharge channel – the thunder cloud may be two kilometres or more above the ground – is rarely heard under such circumstances because the initial bang is literally deafening, albeit temporarily. Sound travels through air at a speed of 0.31 kilometres per second, so it takes roughly three seconds to go a kilometre. A lightning stroke from cloud to ground may be two kilometres or more long, and a cloud-to-cloud stroke can extend several kilometres. Our ears first hear the thunder from the nearest part of the flash and it will take several seconds – more than half a minute in exceptional cases – until the rumbling from more distant parts of the discharge channel finally ceases.

The varying sound of thunder is the result of a variety of factors: the irregular shape of the discharge channel, the existence of several instantaneous discharges at different locations, the branching of 'forked' lightning, and echoing from nearby hills and mountains. When thunder is heard from a distance of, say, three kilometres or more, the higher frequencies are more rapidly attenuated leaving behind a long, low-pitched rolling sound. Dense cloud and heavy precipitation also cause the higher-pitched parts of the thunder to be muffled. This attenuation explains the characteristic loud, high-pitched, crash followed by a relatively quieter, lower-pitched rumble caused by a moderately close strike.

But it is the lightning, in particular the cloud-to-ground (CG) discharge, that is the danger to life and property. There are, on average, two CG strikes per square kilometre per year across most parts of England and Wales. As we noted in the first chapter, the probability of a person being struck and killed by lightning in any year is 13 million to one, but the chance of your house being struck and damaged is much higher, at about 10,000 to one. More often than not, damage is slight, but serious loss may occur if the lightning triggers an electrical fault which in turn might lead to a fire, or if the discharge sets a thatched roof aflame, or if it demolishes a chimney stack.

Severe thunderstorms occur most frequently in London and the Home Counties, the Midlands, East Anglia, Lincolnshire, Yorkshire and Lancashire. Different studies identify different locations as Britain's thunderstorm capital, but the most recent evidence points to a particular hot spot over the Pennines between Manchester, Sheffield and Huddersfield, including Saddleworth Moor.

Until the early 1950s many meteorologists believed that real tornadoes never happened in Britain. Reports in the press were either exaggerations or just figments of the imagination brought on by the stress caused by a severe thunderstorm. Ernest Bilham, the Met Office's senior climatologist at the time, wrote a 350-page book called *The Climate of the British Isles* – the UK's foremost reference volume on the subject until about 1980 – in which the word 'tornado' did not appear once. Even the more credulous Professor Gordon Manley, the doyen of university climate experts, mentioned only four examples in 300 years in his classic book *Climate and the British Scene* published in 1952. Then things began to change. Just over half a century ago, on 21 May 1950, one of the most memorable meteorological events of the twentieth century happened as Britain's longest-lived tornado cut a path some 110km long through the English countryside. Other British tornadoes in the 1950s were successfully photographed and the subject was extensively reviewed by Professor Hubert Lamb who presented a more balanced account of these exceptional phenomena to the Royal Meteorological Society. During the past 35 years the Tornado Research Organization (TORRO) has substantially extended our knowledge of the mechanisms of tornado formation and has also compiled an exhaustive chronology of tornadoes in the British Isles.

According to TORRO there is an average of 20 to 30 days per year when tornadoes occur in the UK, but such a statistic hides the fact that multiple outbreaks occur on some days, often on a vigorous cold front during the autumn or early winter. These apart, the most favoured time of the year is between May and July, and the geographical distribution shows a preference for the Midlands and northern Home Counties; Bedfordshire is said to have the highest observed frequency of tornadoes per unit area in the entire British Isles. Although British tornadoes are, broadly speaking, less intense and less damaging than those that occur in the USA, they can cause serious structural damage to buildings, and they have over the years been responsible for a number of deaths. In recent times many will remember the Moseley-Balsall Heath tornado in Birmingham on 27 July 2005, and the Kensal Rise tornado in north-west London on 7 December 2006.

Hail is not normally a great threat in the UK. Hailstorms are typically short lived, affect a very restricted area, and only rarely are hailstones bigger than, say, the size of gooseberries experienced in any one place. Once or twice a year, hailstones as big as plums or golfballs may be reported. Chunks of ice this size can break windows, damage cars, ruin greenhouses, kill birds and small animals, and strip trees of their leaves, especially when they are associated with strong or gusty winds.

Averaged over a long period, the frequency of hail varies from 20 to 25 days per year in northern and western Scotland and northwest England, to just five to ten days per year in eastern counties of England. Winter is the favoured season in northern and western regions, but the spring months bring the most frequent reports of hail in eastern, central and southern England. But these winter and spring reports of hail usually involve very small hailstones, whereas large hail is more likely in high summer. On average, in July and August hail occurs on about one day in 50 in most parts of the UK, and it is only marginally more common in June and September, and these are the four months when destructive hailstorms are most likely to happen. On rare occasions hailstones may fuse together to form large irregular chunks of ice. The largest known British hailstone fell during a day of widespread thunderstorms on 5 September 1958; it was picked up at Horsham in Sussex, it weighed roughly 140 grams (five ounces), and it was over ten centimetres in diameter.

Flash floods are also rare in the UK, but they are a bigger threat to life and property than any of the other rapid-onset phenomena. First of all, though, we should be clear what we mean by the term, because the phrase 'flash flood' is widely misused by the news media (and even by some weather forecasters). A flash flood is a violent phenomenon that occurs almost instantaneously, typically in the form of a wall of water moving down a valley at great speed, and usually triggered by a violent rainstorm somewhere in the upper reaches of the valley; sometimes there is no rain at all at the place where the flood hits. It can roll large boulders, tear out trees, destroy buildings and bridges, and scour out new channels. The term is a relatively new one, having been in use in the USA since the 1930s, and in the UK since the mid-1980s. Its newness and the attractiveness of the alliteration have resulted in overuse of the expression to describe any old flood – a journalistic weakness known in the trade as 'adjectival inflation'. Weather forecasters, though, should know better.

Destructive flash floods in Britain involve a collusion of meteorology and geography. A heavy downpour amounting to at least 75mm of rain in two hours or less will cause short-lived flooding in most parts of the country, but if the cloudburst occurs over a relatively small, steep river catchment then the flooding can be catastrophic. Rainfall of this intensity is largely confined to the months between May and October, and is perhaps more likely in south-west England than in any other part of Britain. South-west England is also characterized by small, steep river catchments. A typical West Country catchment will respond to heavy rain very quickly, delivering water from the headwaters to the sea in a matter of hours; if this happens while the rain is still hammering down, the flooding is made even worse. One needs only mention Lynmouth in August 1952 and Boscastle in August 2004 to remind people what sort of elemental destructive power is unleashed during a true flash flood.

24-hour hazards

The next range of hazards includes those which develop over an hour or two or more, but which usually pass within 24 hours. Once again the repercussions may go on for weeks or months. The damage-to-time ratio is high, though clearly not as high as for rapid-onset hazards; nevertheless these events are mostly very newsworthy. In our part of the world they will include severe gales, coastal flooding and major snowstorms.

Severe gales have probably been responsible for more deaths in the British Isles over the centuries than any other sort of weather, and that is as true of the early twenty-first century as it was in Victorian, medieval or even Roman times. The chronology stretches from Julius Caesar's abortive attempts to cross the English Channel in a north-easterly gale in 55BC to a fierce westerly gale on 18 January 2007 which took nineteen lives. Gale-related deaths are reported in most years, and damaging winds occur routinely nearly every autumn and winter and occasionally at other seasons too. The maritime character of the peoples of the British Isles means that there is nearly always a good deal of activity – economic, military, recreational – going on in our coastal waters, and over the years a substantial proportion of the losses have occurred there. Nor should it be forgotten that the establishment of the UK's national meteorological service, now called the Met Office, was triggered by one particular storm at sea, way back in October 1854, and the first warnings and forecasts were developed specifically with marine activity in mind.

Measuring windspeed presented one of the biggest challenges to early instrument makers who encountered four different kinds of problem: first, the wind can be measured as a velocity (including both direction and speed) or as a force exerted on stationary objects; second, any instrument has to point into the wind in order to measure velocity or force; third, the wind comprises an endless succession of gusts and lulls so it was necessary to measure both the average sustained speed and the speed of the gusts; and fourth, frictional and other effects mean that wind speed increases (and variability decreases) with height above the ground.

Once the basic technical issues had been sorted out, the biggest problem meteorologists had was in deciding where to site their anemometers. We, of course, live at the very bottom of the atmosphere and most of our houses, gardens, offices, factories and transport systems are in the lowest 20 metres or so, but this is precisely the area where winds are reduced and distorted by friction and turbulence. Friction reduces average wind speed at head height by more than 75 per cent when compared with the wind in the free atmosphere 50 metres or so above us. And one only needs to walk the streets in any British town on a windy day to appreciate the degree of turbulence and funnelling caused by buildings. The accepted standard for anemometer sites is thus a compromise: high enough to escape much of the distortion caused by buildings, trees, and other obstructions (what meteorologists called 'surface roughness'), but low enough to be relevant to the very shallow layer of the atmosphere in which we conduct our everyday lives. Thus most instruments are sited at 10 metres above their surroundings. Even so, in a city environment the residual turbulence will still result in a gust ratio of more than two – that is, the strongest gusts are more than twice the average wind speed. One other important aspect of wind, particularly significant when relating wind speed to gale damage, is the fact that the wind speed is measured at a point but damage is caused by the force exerted by the wind over an area. Thus gale damage is roughly proportional to the square of the wind speed. In practice, the degree of damage will be exacerbated the longer the gale persists.

Snow is one of the most variable aspects of the British climate – variable both in space and in time. During the standard reference period 1971–2000, snow or sleet was observed to fall on just two days per year on average in the Isles of Scilly, twelve days in central London, sixteen in Manchester, twenty in Edinburgh, 35 in Aberdeen, more than 50 at Aviemore and Braemar, and an estimated

170 days on the summit of Ben Nevis. Snow covers the ground in the Isles of Scilly on an average of one day every eight years, on three days per year in central London, nine in Manchester, fifteen in Edinburgh and Aberdeen, 65 at Braemar and approximately 200 days on Ben Nevis. If we consider low-lying districts only, there is a trend from south-west to north-east, with the coasts of Cornwall, south Devon and Pembrokeshire often snow free, while snow is also comparatively rare on other west-facing coasts such as those of Lancashire, Galloway, Ayrshire and Argyll. People who live along the North Sea coastline are much more likely to wake up to a snow cover, with the highest probability in north-east Scotland. However, the main influence on the frequency of snow cover is altitude. Very roughly, there is an increase of one day per winter for every 15 metres (50 feet) above sea-level. Thus Plymouth has an average of one day with snow on the ground while Princetown on Dartmoor has 28 such days; central London has three, but Hampstead Heath has thirteen; Manchester has nine but the eastern outskirts of Oldham and Rochdale have 25 and Saddleworth Moor has 40.

There is also a huge contrast in the frequency of snow cover from year to year. Except on southern and western coasts, completely snowless winters are uncommon. In the London suburbs, for example, there was no snow on the ground in only five of the last 50 winters. In Birmingham there has been just one snow-free winter in the last half-century, but the majority of years have fewer than ten days with a snow cover; by contrast, the severe winter of 1962–63 had 75 such days while those of 1978–79 and 1946–47 had 60 each. It is also clear that the frequency of snowfall is extremely sensitive to climate change: in the English Midlands, for instance, the number of days per winter with snow lying has declined from an average of fifteen days between 1938 and 1987 to just six days between 1988 and 2007, while the rise in mean winter temperature between the two periods amounts to just 0.9°C. That also means that the frequency of major snowstorms is also decreasing and, assuming that the underlying warming trend in our climate continues, they will become even rarer in the coming decades. However, the meteorological circumstances which favour heavy snowfall still do occur from time to time, and will continue to do so for many years yet, so we should not run away with the idea that seriously disruptive snowstorms are no longer possible in the UK in the warmer world of the twenty-first century. Furthermore, as we noted earlier, the declining frequency of a particular hazard usually means that we will become less able to cope with more modest examples of the hazard in the future. The historically unremarkable snowfalls of late January 2003 (the night of the 'M11 gridlock') and early February 2007 provide excellent illustrations of that.

Coastal flooding is an example of the extremes in two different spheres – in this case meteorological and oceanographic – coming together to create an especially nasty form of hazard. A powerful gale from the right direction, especially if prolonged, will cause high seas around Britain's coastline, but it requires the approximate coincidence of the gale with a high astronomical tide (that is, a spring tide) to overtop sea defences. As a matter of fact, one might identify third, fourth and fifth facets which influence the vulnerability of parts of our coastline to this particular hazard: geography, because much of the coastline of England and Wales is sinking relative to the sea and as a consequence older sea defences very gradually become less effective; socio-economic, because planning decisions to allow new developments to take place in low-lying coastal districts expose more and more people to the threat; and political, because governments periodically exercise a benign neglect over our coastal defences for economic reasons and this, too, increasingly exposes residential and commercial developments to the threat of sudden and catastrophic flooding.

This was indeed the policy of successive British governments during the 1930s and 1940s, but neglect mutated into negligence in a matter of hours on the night of 31 January to 1 February 1953 when the most catastrophic of storm surges swept down the North Sea and flooded some 750 square kilometres of land, from the East Riding of Yorkshire southwards to Kent. Over 300 people died in the storms that night. Winston Churchill, Prime Minister at the time, promised that the sea would never again be allowed to launch such an effective and dramatic invasion, and his government initiated a rolling programme of improvements to Britain's sea defences, culminating in the completion of the Thames Barrier in 1982. These improvements saved many parts of eastern England from a potentially worse disaster in January 1978 when another violent northerly gale coincided with a spring tide. Churchill's policy was overturned in 1993 by John Major's government whose environment and agriculture departments questioned the cost of defending agricultural land in an era of over-production, and there have been a number of examples of 'managed retreat', notably in Essex, Norfolk and Lincolnshire, since then. Such a policy has the added advantages of returning sections of our coastline to a more natural character with extensive salt-marshes, dunes and lagoons, and this in turn makes the coastline more absorbent when flooding does occur. It also allows central and local government to concentrate resources on maintaining and improving the defences around residential and commercial developments. However, the changing climate is expected to induce an accelerated rise in sea level later this century and during the next, and that could well trigger further, more dramatic policy changes in respect of Britain's coastal defences.

Seven-day hazards

More gradual in their onset, and often longer lasting, is another range of hazards, including fogs and smogs, heatwaves and river floods. These can develop so slowly that the true measure of the threat may be concealed for some time, perhaps for days. Although less dramatic in their development than their rapid-onset cousins, these hazards may last a week or more, and it is their protracted nature that inflates the costs to the national or local economy, and to the health and ultimately the lives of the people.

River flooding, like coastal flooding, is often exacerbated by human activity. Since the British Isles emerged from the last Ice Age, flooding has been an integral part of our natural environment. It is, no doubt, scant comfort to the inhabitants of towns like Carlisle and Shrewsbury and York which routinely experience inundation to be told that their disaster was entirely normal. But there are so many uninformed commentators who always link floods with climate change that it is essential to explain that there is no need to look for any 'special' reason for them. Indeed, being able to blame, say, global warming actually provides an excuse for all those agencies and authorities whose job it is to mitigate this particular type of hazard. Having said all that, the impact of flooding on our homes and communities is much greater now than it was, say, 50 or 100 years ago. For this we should be able to hold to account those in authority for the planning (and other) decisions that they have made over the years: culverting streams, straightening water-courses, draining marshes and water meadows, and above all, building new roads, houses and business premises on flood plains. All these activities contribute to the severity of flooding along our rivers, but they also expose a larger number of people and a bigger slice of the nation's wealth to the hazard. Needless to say, all these activities have become endemic in the last half-century. Thus the recent inflation in weather-related insurance payouts is not a consequence of an increased frequency of severe weather; rather, it is a reflection of how well insured we now are, how effectively we keep our insurance cover up to date, and how much more costly our property and belongings are to repair or replace. It also tells us a lot about the irresponsible way in which planning permission applications which are refused by planning departments are overturned by local councillors, allowing property developers to build new homes, schools, supermarkets and other businesses on the flood plains of rivers. Flood plains are there for a reason – to permit periodic flooding when rivers overtop their banks – and they are flat and easy to build on because of the silt and mud which have been deposited there at regular intervals over aeons of time. We all knew this once, and, in times past, flood plains were never built upon. But since the 1950s and especially in the last 30 years, greed has begotten amnesia. The severe and repeated bouts of river flooding which occurred in Britain between October 2000 and April 2001 have led to a rethink, and the fact that most insurance companies will now not offer cover against flooding to new properties built on flood plains will impose a severe financial penalty on such developments in the future.

Smog episodes have all but vanished from the UK following periodic bouts of clean-air legislation throughout the last 50-odd years, beginning with the Clean Air Acts which followed the great London smog of December 1952. There had been plenty of attempts to introduce effective laws before the 1950s, but they were always withdrawn or diluted when they came up against the vested interests of commerce and industry. It required a catastrophe of the scale of the 1952 smog when an estimated 7,000 people died prematurely to overturn those powerful interests; it is estimated that approaching 100,000 people died before their time in the UK as a consequence of illnesses exacerbated by concentrated air pollution during periods of heavy industrial smog during the nineteenth and twentieth centuries. Serious fogs can still happen, as travellers trying to leave the UK just before Christmas 2007 found to their cost, but they, too, have become less frequent during recent decades. Now that the risk to our health has diminished, the main threats from thick and/or persistent fog are twofold: the enhanced risk of accidents, chiefly on rail and road, and the economic losses resulting from traffic congestion on roads, in coastal waters and at airports. Public education and the installation of motorway lighting in many areas have contributed to a sharp decline in multiple pile-ups (so-called 'motorway madness') in fog since the early 1970s when over 1,000 vehicles were involved in such multi-vehicle accidents on the then rudimentary motorway network over a three-year period.

It would have been laughable in the 1950s and 1960s to suggest that a heatwave was a hazard. In only ten of the 25 years from 1950 to 1974 did the temperature exceed 32°C (90°F) anywhere in the UK, and never was that threshold passed on more than two consecutive days. The hot summers of 1975 and 1976 changed our perception of what British summers could do, and the increasing frequency and longevity of heatwaves from 1989 onwards have provided us with a powerful warning of how unpleasant and life-threatening the summers of the middle and latter parts of the twenty-first century might become. It is a warning we ignore at our peril. The Department of Health recognizes five categories of people at particular risk during extended periods of hot weather, including the very old (especially those living alone or in care), those with a mental incapacity, the bed-bound, those taking certain kinds of medication, and the very young (under four years old). The risk is increased further for people who live in inner cities, for those in top-floor flats, and anyone whose place of work is hot anyway, such as a foundry or a bakery. These groups are therefore most likely to suffer heat exhaustion which, if untreated, will eventually lead to heatstroke. When both temperature and humidity are high, our bodies work very hard to keep our internal temperature stable: we perspire heavily, and it is advisable to keep physical exertion to a minimum, to wear loose, light-coloured clothing, to seek shade, to drink water, and to attempt to create air movement by using fans. If these mechanisms fail, our core body temperature rises and heatstroke follows; unless immediate steps are taken to counter the overheating, heatstroke will be fatal. Sunstroke is similarly perilous. When high temperature is accompanied by exceptionally low humidity we perspire readily but we are less aware of it because the perspiration quickly evaporates. Unless we make a conscious effort to maintain body fluid and natural salts by frequent drinks of a suitable nature we quickly become dehydrated and disorientated, sunstroke follows and this too may be fatal. Alcoholic drinks are not a good idea: alcohol absorbs water, the natural cooling mechanism of our bodies becomes less effective, and this may accelerate the onset of both heatstroke and sunstroke.

The Civil Contingencies Secretariat, in conjunction with the Department of Health, has introduced a heatwave plan in order to provide advice through the media, to help identify at-risk groups, and to encourage additional help through the voluntary sector. The plan provides for four levels of warning: 'Awareness', which is a sort of early warning; 'Alert', which provides a warning of an imminent spell of very high temperatures likely to affect people's health; 'Heatwave', which is aimed at maximizing information and help while the hot spell is actually with us; and 'Emergency', which will kick in only on those very rare occasions when the heat is so extreme or protracted that the integrity of the nation's infrastructure begins to break down – for instance, hospital emergency departments overwhelmed, power failures, severe restrictions in the water supply. Such a situation might arise if the French experience of August 2003 ever were to happen on this side of the English Channel. The trigger temperatures for most regions of the UK, by the way, are afternoon maxima of 30°C or more and overnight minima of 15°C or more, for three consecutive days; in London and the south-east the thresholds are a degree or so higher. For anyone who thinks that this is the nanny state gone mad, it is worth reminding them of the Europe-wide death toll during the summer of 2003, when an estimated 30,000 perished from a heat-related cause, of which 15,000 were in France alone, and approximately 2,000 in the UK. During the summer of 2006, another hot one, the 'Alert' category was reached on twelve days, the 'Heatwave' category on seven days, and an estimated 700 heat-related deaths were reported. These numbers are necessarily approximations because they are calculated by comparing the weekly death rate with the average death rate during the equivalent week in earlier years, but it should also be made clear that these are deaths over and above the normal mortality rate; they are not people who would have died anyway a few weeks later.

Slow-onset hazards

So far, we have considered hazards which have had a single dominant meteorological cause. At longer time-scales we have to consider abnormal seasons when particular weather patterns keep repeating themselves, abnormal years when some external influence such as a major volcanic eruption may exercise control over the weather of the British Isles and other parts of western Europe, and abnormal decades when, for instance, fluctuations in ocean currents in the Atlantic may bring about a radical – if temporary – change in our climate. This last example shades imperceptibly into the slowest and longest-lasting hazard of all, that of climate change.

Abnormal seasons, dominated for long periods by a particular kind of weather, are surprisingly common in the UK. In our relatively friendly and equable climate they may not pose much of a threat to human life, but they can be costly in an economic sense, and they can also be extremely inconvenient to the smooth running of our day-to-day lives. Cold and snowy winters give rise to repeated congestion and delay in our transport infrastructure, raise the cost of repairing roads damaged by frost, increase the number of people in accident and emergency hospital departments thanks to broken limbs following accidents on icy footpaths, and inflate the proportion of working time lost through illness, accident and inability to get to work. Dry springs and summers necessitate increased irrigation of farmland, reduce crop yields, threaten water supplies and put heath and moorland at risk from fires. Cold, wet summers also reduce the yield of many crops, hinder harvesting and are detrimental to the financial success of our coastal resorts, summer sports and other outdoor attractions. Our changing climate seems to have put cold winters out of business (although that does not mean to say that next winter could not be a bad one), but during the last twenty years alone we might identify the long hot summers of 1989, 1990, 1995 and 2003, the excessively wet autumns and/or winters of 1989, 1989–90 and 2000–1, the wet summers of 1998, 2002 and especially 2007, and the droughts of 1989–92, 1995–97, 2003 and 2004–6, all of which had a measurable impact on the British economy, and also on our everyday lives.

Droughts are a serious but routine hazard in the UK. It may seem odd that a country recognized throughout the world for its regular and copious rainfall should allow itself to suffer from such a problem, but it all comes back to what we are used to. If our rainfall is normally sufficient to provide adequate water – with appreciable wastage – to our densely populated island, then any shortage of rain over an extended period will cause problems. Ally that to a growing tendency towards profligacy, to an expanding population in the driest part of the country, and to the privatization of the water-supply industry under a weak regulatory authority, and you have trouble.

Drought can be defined in many different ways, depending on your particular concerns. Most simply, a *meteorological drought* may occur when rainfall during a specified period falls below a particular threshold. For instance, in Britain this may be represented by a 50 per cent deficit over three months, or a 15 per cent shortfall over two years; the actual figures may be selected to demonstrate a particular point. A more useful analysis involves calculating 'return periods' for a range of time intervals at individual locations; for instance, a 20 per cent shortfall over twelve months in London has a return period of roughly fifteen years – in other words it will recur, on average, once every fifteen years. Arguably the most outstanding meteorological drought in the British climatological archives happened in 1893. That year delivered an exceptional spring of warmth and sunshine, and rain fell on only a handful of days. The dry weather set in during the first week of March and lasted until early July. In the London area there was only 35 per cent of the normal rainfall during the four months from March to June inclusive; at Mile End in east London no measurable rain at all fell between 4 March and 15 May, a period of 73 consecutive days, while Twickenham in Middlesex reported 72 successive days with no rain.

An *agricultural drought* happens during the spring and summer – the growing season – when evaporation normally exceeds rainfall anyway. During a settled year when the weather charts are dominated by high pressure systems, rainfall is low while evaporation is higher than average thanks to increased sunshine and rising temperatures. Growing plants suffer stress caused by a shortage of moisture and nutrients, and this leads to reduced growth and lower crop yields; if the moisture deficit continues indefinitely the plants will ultimately die. Desiccated meadows and grassland also present a grave problem for husbandry: animals have to be fed from stocks of hay and silage, or from concentrated feed, and this in turn depletes feeding resources for the following winter. Milk yields are reduced, while sheep and beef cattle take longer to fatten up for slaughter. Prices of home-grown fresh vegetables and fruit soar at the same time as the quality deteriorates. Away from the farms, woodlands, moors and heaths are at a much-increased risk of fire, whether set by humans either accidentally or deliberately, or by sunlight focused through discarded glassware, or by lightning strikes. Such fires are costly in a variety of ways, damaging commercial stands of trees, occupying firefighters, threatening property, and damaging the local flora and fauna. The twentieth century included some prime examples of agricultural droughts, notably in 1921 (the driest calendar year in 300 years of records over a wide area), 1959 (the driest February to September on record) and of course 1976 (the driest April to August on record).

The water-supply industry requires adequate winter rain to make sure reservoirs are full and ground water is topped up. Thus a *hydrological drought* might be said to occur when winter rainfall fails to counteract the soil-moisture deficit from the previous summer, while reservoirs are insufficiently replenished. This sort of drought can extend over a period of several years even though intervening summers may be rather damp. Such droughts were extremely serious before the piped water network for towns and cities was developed in the late nineteenth and early twentieth centuries. The rapid growth of towns during the opening decades of the nineteenth century concentrated millions of people in urban areas where the only supply of water was from wells and pumps. A sequence of very dry years in the 1850s led to the failure of thousands of wells and springs, water had to be transported into urban areas on carts, and the threat to health and life became very grave indeed. (Statistics for deaths attributable to the drought during this period are simply not available, but one might speculate that they were in the tens of thousands.) Drying wells are also more susceptible to disease, and cholera, especially, was rife in hot, dry summers. During the 100 years or so up to 1988, water supply was largely in the hands of regional water authorities although a few heavily regulated private companies also existed. Though not always terribly efficient, the overriding priority of these organizations was to maintain an adequate water supply for their customers. Since privatization in 1988, and the marked relaxation of supervisory regulation which happened at the same time, the prime responsibility of the water-supply companies is to their shareholders, while customers seem sometimes to be treated as little more than a nuisance. This was most dramatically illustrated in 1995 when Yorkshire Water failed in their duty to provide an adequate supply of water following a mere five months of dry weather. The drought of 1975–76 was undoubtedly the most severe of all hydrological (as well as agricultural) droughts since records began, and more recently an extended dry period between November 2004 and September 2006 led to a great number of threats of serious water shortages, although most of the threats failed to materialize.

Natural and anthropogenic climate change over decades and centuries represent the extreme end of the range of hazards which began with near-instantaneous phenomena like lightning and tornadoes. Our changing climate will also change the nature, frequency and intensity of the shorter-period hazards that we have already discussed. For although it is wrong to point the finger at every bit of unusual weather and blame it on global warming, we have to recognize that an underlying warming trend will result in a changed background against which rapid-onset hazards develop. Thus we are already noticing a decline in the frequency of disruptive snowstorms, fewer seasons of extreme cold, but a rise in the number of killer heatwaves. Whether the frequency of tornadoes, severe gales, floods and droughts also changes is, in spite of much uninformed comment, not at all clear. We shall return to this topic in a later chapter.

Chapter 4

Snowstorms: two case studies

Introductory remarks

The next five chapters look more closely at some of the more extreme and newsworthy weather disasters in the historical record. These case studies are arranged in pairs, with the illustrations taken from different eras, sometimes more than 100 years apart. This enables the reader to compare and contrast the different impacts that similar events have had on the country and on individual communities in different times. Care has been taken to avoid revisiting many of the best-known events – the great storm of October 1987, the Lynmouth tragedy of 1952, the historic drought of 1976, and the North Sea floods of 1953 – which have been covered in great detail in numerous other publications over the years. Nonetheless, these historically important events take their proper place in the chronology of disasters which follows in Part 2.

The snowstorm of 18–21 January 1881

There had been nothing like it for two generations, nothing to compare since before Queen Victoria came to the throne. Some said it was as bad as 1836; those with better memories suggested that the quantity of snow that fell was similar but that the ferocious easterly gale and coastal flooding of January 1881 were far, far worse than the earlier instance. The very oldest men and women in towns and villages across southern England muttered darkly about January 1814, but mention of even that distant date revived memories of a massive snowfall spread over several days but without the additional ravages caused by the wind. Looking back from our vantage point, this absence of paralysing snowstorms for several decades might seem surprising, because the winters of the nineteenth century were routinely much colder and snowier than anything we experience today. The coldest winter of the past two decades fell in 1995–96 when the Central England Temperature (CET) for the three months of the winter quarter was 3°C; there were 32 colder winters between 1801 and 1900. The twentieth century had its fair share of severe and disruptive snowstorms – April 1908, Christmas 1927, several in the 1940s, February 1978 and January 1982 immediately spring to mind – but for the combination of snow and wind none of these quite matched the storm of 1881.

So, as the cliché has it, not even the oldest inhabitant had seen the like. And yet a similar snowstorm struck the same part of the UK just ten years later, in March 1891. On that occasion Cornwall and Devon were hit more severely than in 1881, the Home Counties rather less so. The story of the later storm has been told many times, nowhere better than in Clive Carter's book *The Blizzard of '91*, but the background to the earlier event deserves to be examined as closely, for, in the view of many experts, it was meteorologically the more interesting of the two.

The period between 1878 and 1895 was the coldest since the years following the cataclysmic eruption of Mount Tambora in what is now Indonesia in April 1815 (the following year, 1816, was widely known in the northern hemisphere as 'the year without a summer') and most winters during the more recent period contained at least one month with a CET close to or below 2°C. The winter of 1880–81 began quite benignly, and for much of December there was little frost or snow thanks to persistent south-westerly winds, although it was a wet month (see Table 4.1).

The barometer rose strongly during the opening days of the new year, ending the sequence of very wet days and settling the weather down, especially in the southern half of the country, but it remained relatively mild for the time being practically nationwide. However, the high-pressure system which had formed over southern Britain gradually moved northwards, and by the 5th it lay over Scotland, allowing a stiff nor'easter to develop over England and Wales, and this increasingly sharp wind introduced gradually lower temperatures and some light snow flurries over the subsequent five days. There now followed a week of intensely cold weather with severe

Table 4.1 Temperature for central England, rainfall and sunshine averaged
over England and Wales

Month	CET	Difference from 1971–2000 average	Rainfall	Percentage of average	Sunshine	Percentage of average
December	5.1°C	0.0°C	123mm	122%	38 hrs	80%
January	−1.5°C	−5.7°C	36mm	38%	54 hrs	98%
February	3.2°C	−1.0°C	110mm	165%	48 hrs	63%

night frosts, daytime temperatures close to or somewhat below zero, and there were also some heavier and more widespread falls of snow as a small depression travelled southwards down the eastern side of the country between the 11th and 13th, and again in Scotland on the 16th and 17th. By now, the snow was 30cm deep at Wick in Caithness, 10–15cm deep around Aberdeen and over much of north-east England, but it lay no more than 5cm thick elsewhere. The temperature fell below −10°C quite widely every night for three weeks from the 6th/7th, and below −20°C overnight 15th/16th, 16th/17th and 17th/18th; lowest readings included −22.2°C and −26.7°C at Kelso in the Scottish Borders on the latter two nights, and these remain the lowest readings ever recorded on these particular dates in the UK. At York the maximum daytime temperatures on the 17th and 18th were respectively −6.1°C and −7.8°C, while the mercury failed to rise above freezing point over a large part of the English Midlands for fifteen consecutive days from the 12th to the 26th inclusive. By the 18th the ground had become as hard as iron, frozen to a depth of 10–15cm over much of Britain; only in Cornwall and Devon had daytime temperatures climbed much above zero and even here there were sharp frosts at night.

Thus was the scene set for the main performance. The Meteorological Office, now under the directorship of Robert H. Scott, had been issuing a Daily Weather Report since 1862, and this publication now included a rudimentary synoptic chart, a temperature chart, twice-daily observations from 30 sites around the British Isles and twenty more in other parts of western Europe, a third observation from a handful of stations, a manuscript assessment of synoptic developments during the preceding 24 hours, and some elementary regional weather forecasts. The weather charts which were drawn in those days were, by modern standards, extremely basic, but, at least to my knowledge, no detailed re-analysis has even been undertaken until now. No upper-air observations, of course, are available, but a re-drawing of the isobars from all available surface observations has been undertaken with the addition of fronts; this allows a notional mid-tropospheric circulation pattern to be inferred which in turn enables some 'tweaking' of the surface analysis. This combination of actual weather reports and inference based on later developments in the science of synoptic analysis provides a more detailed and comprehensive examination of the great snowstorm of January 1881 than has previously been attempted. It also helps to explain why the snowfall was so long lasting in southern counties of England and why quite so much snow accumulated in that part of southern England between south Devon and the Isle of Wight.

Weather observers noted a brisk fall in barometric pressure over Ireland, Wales, south-west England and western France during the evening of the 16th, and on the 17th a continued fall of pressure in the Southwest Approaches contrasted with a steady rise over Scotland, Denmark and especially Scandinavia. Thus the gradient for easterly winds steepened sharply and by that evening the wind was blowing at Beaufort force 6 or 7 over southern Ireland, south Wales and southern England, and at force 8 in Brittany at both Brest and Lorient. Snow had already started to fall at Lorient.

The original analysis of the event described it simply as a deep depression moving slowly along the English Channel, and over the subsequent century and a quarter nearly all references to this particular snowstorm have repeated those words. Other minor depressions appeared to form and dissipate almost at random, but the re-analysis shows that these secondary features represent important elements of a rather more complex sequence of events. On the 17th, an old shallow depression, probably coincident with a cold-air 'low' in the upper atmosphere, was slow moving just south of Ireland, while a polar depression centred north of Shetland was heading in a south-westerly direction. An active depression travelled from near the Azores in a north-easterly direction, tracking past the north-west corner of Spain that afternoon, thence turning slightly to the left and deepening abruptly, to be located near Guernsey with a central pressure of 972 millibars by 0800 on the 18th, and in mid-Channel between Cherbourg and Portland, still at 972 millibars by 2pm. Until now the depression had been steered, at least in part, by a south-south-westerly flow in the upper troposphere, but its deceleration during the morning of the 18th indicates that the steering flow was weakening, which in turn suggests that our depression had now developed its own circulation aloft. The surface depression now seemed to follow a tight circular track, presumably around the upper-air centre, before resuming its north-eastward passage. Thus the centre of the low lay just inland of the Dorset coast at 974 millibars at 6pm on the 18th with a south-westerly wind at Hurst Castle, at the entrance to the Solent, and a north-westerly one at Prawle Point, the southernmost tip of Devon; everywhere else in England and Wales an easterly gale was raging.

No further observations were taken until 8am on the 19th, and by then the depression appears to have travelled slowly west then south-west into Lyme Bay – Prawle Point still northerly, Hurst Castle now easterly – with a central pressure close to 980 millibars. From here it turned south-east towards Cherbourg by 2pm having filled to 983 millibars, and then resumed a north-easterly track to be mid-Channel between Dieppe and Newhaven by 6pm, still filling, at 986 millibars. By the morning of the 20th the centre lay over Belgium, having crossed the French coast some distance south of Cap Gris Nez. The shallow depression west of Ireland became caught up in the circulation of the Channel storm, crossing the Bay of Biscay and France during the next two days. Meanwhile the polar depression lay off north-west Ireland on the morning of the 19th at 1004 millibars, whence it turned south-eastwards to be over the Irish Sea at 1007 millibars by 8am on the 20th, thence to the Bristol Channel at 1010 millibars by 6pm that day, and to a position off south Devon at 1013 millibars by 8am on the 21st.

By the morning of the 18th the snow had spread northwards from Brittany to reach all southern counties of England from Cornwall to Kent, it was snowing in London and Oxford by early afternoon, and as far north as mid-Wales, Birmingham and Norwich by evening; the easterly wind was now blowing a full gale even inland, and was reported as severe gale force 9 or 10 along the east coast of England. On the 19th the snow gradually retreated towards south-east England, finally clearing from East Anglia and Kent during the evening, and the high winds steadily subsided. With temperatures remaining well below zero in all inland areas, the snow was dry and powdery, and the gale force wind had swept it into massive drifts, even in the streets of London. That seemed to be the end of it all, but there was a nasty sting in the tail. As the polar depression moved southwards across the English Channel overnight 20th/21st, south-easterly winds developed on its eastern flank over mid-Channel, and a convergence line developed for a time across south Hampshire and south Dorset, marking the boundary between moist south-easterlies from the Channel and cold, dry north-easterlies over inland parts of England. Following 30–36 hours of continuous snow during the main event, this convergence line was responsible for up to twelve hours of renewed heavy snow over south-west Sussex, south Hampshire, the Isle of Wight, and parts of Dorset.

Plate 4.1 An engraving depicting the horrendous conditions in Fleet Street, London, during the snowstorm of 18–19 January 1881
(Mary Evans/*Illustrated London News*)

Had there been no snow at all, the easterly gale would have been remarkable in its own right – one of the strongest gales from that direction for which we have records. Thousands of trees were uprooted, notably in East Anglia, roofs and chimney stacks were severely damaged in many parts of England but especially in the London area, and the high winds drove an already high spring tide up the Thames estuary and into the heart of London. Here, river defences were overtopped, and the icy waters flooded into low-lying districts especially south of the river in Deptford, Bermondsey and Southwark. Thousands of people were displaced from their homes, and approximately 100 of them were drowned in basements. A further 100 may have died as a direct consequence of the snowstorm, and the extreme cold resulted in an increase in mortality in London alone of 410, chiefly from bronchitis and pneumonia. By factoring up this rate to the whole of the UK, we may estimate a death toll from the cold and snowy weather of January 1881 to have been of the order of 3,000. Wildlife was also badly hit, and the large numbers of birds that perished was remarked upon by many contributors to the meteorological journals.

Road and rail transport were completely dislocated by the heavily drifted snow, and as a consequence food, fuel and other supplies failed to reach many towns and villages. Plymouth was without a water supply for almost a week as a result of the aqueduct carrying the water from Dartmoor to the city becoming frozen and choked with snow; it took 1,000 men five days to restore the supply. Communication by telegraph was also disrupted, with thousands of lines brought down by the strength of the wind and the weight of ice, while postal services were suspended for over a week over much of southern England. On the railway network, many locomotive engines were completely buried by the snow in the deeper cuttings, and entire trains were stranded in drifts for more than a day. The Tring cutting, in Hertfordshire, was eventually cleared by filling 1,700 trucks with snow. Three trains were snowed in on the stretch of line between Reading and Oxford alone, and the passengers on one of those services were finally rescued after 26 hours; the train itself could not be moved for almost five days. In Wessex and the West Country most railway lines were completely blocked for three or four days, and in Devon the Holsworthy and Ilfracombe lines remained impassable for seven and eight days respectively. The

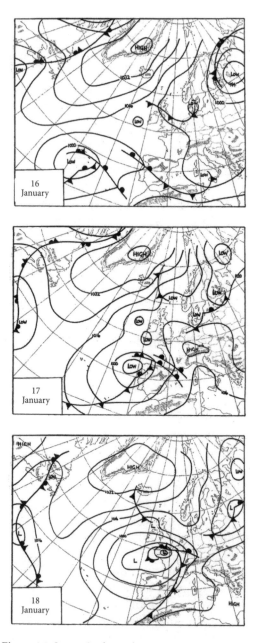

Figure 4.1 Synoptic charts for 16–21 January 1881

Figure 4.1 continued

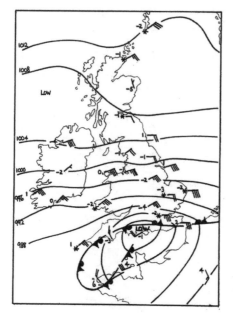

Figure 4.2a Detailed weather chart for the British Isles, 18 January 1881, 6pm

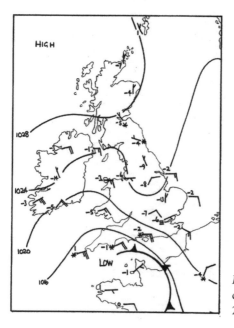

Figure 4.2b Detailed weather chart for the British Isles, 21 January 1881, 8am

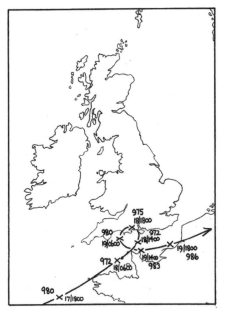

Figure 4.3a Depression track, 17–19 January 1881

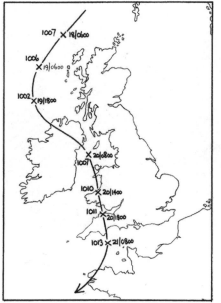

Figure 4.3b Depression track, 18–21 January 1881

Great Western Railway reported that 51 separate passenger trains and thirteen goods trains were stranded at the height of the snow-storm, while the Great Eastern Railway had eleven trains stuck in snow in Norfolk, two of them for three days.

The largest accumulations of snow from the main snowstorm were in the Isle of Wight and adjacent parts of south Hampshire and south-east Dorset, but these were exactly the areas where snow fell heavily again during the night and early morning of the 20th/21st. At Newport the first fall averaged 40cm, and the second 45cm, resulting in a total level depth of 85cm. At St Lawrence in the extreme south of the Isle of Wight the total fall amounted to 55cm, while Ventnor and Osborne both reported 45cm. At Ryde, the final depth approached 60cm; one report described the town thus: 'Many of the roads are said to be filled nearly half way up the lamp-posts, and the lamps were left alight the greater part of the day on account of the difficulty getting to them . . . no shops were open, the drifts in some cases being above the tops of the doors and shutters.' The Ryde District railway line was not re-opened until the 26th, eight days

Plate 4.2 An engraving showing passengers leaving a train which had been stranded for several hours at an unnamed railway station in southern England during the snowstorm of 18–19 January 1881
(Mary Evans/*Illustrated London News*)

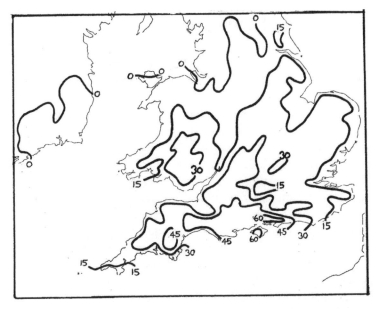

Figure 4.4 Maximum level snow depth (in centimetres), 18–21 January 1881

after the first snowstorm. At Shorwell, south-west of Newport, the school house was completely buried, and a tunnel had to be dug from the road to the front door to rescue the schoolmaster and his family. The village of Chale had no bread for seven days. On the mainland the area around Portsmouth was particularly badly hit with upwards of 60cm of snow to the north of the city in the vicinity of Portsdown Hill, while drifts here were reported to be five metres high. A meeting hall in Portsmouth itself was demolished when the weight of snow and ice caused the roof to collapse.

In the wake of the snowstorms, the cold intensified for a week, and overnight temperatures fell below –15°C even in southern England. On the morning of the 25th a reading of –21.7°C was obtained at Bury St Edmunds in Suffolk, and on the 26th a minimum value of –22.8°C was logged at Sorrel Sykes, near Aysgarth in the North Riding of Yorkshire, and –22.2°C at Raby Castle just over the border in County Durham. It was not until the 28th and 29th that freshening southerly and south-westerly winds finally lifted temperatures above freezing and initiated a thaw, slow at first, but increasingly rapid as January drew to a close.

The snowstorm of 7–9 December 1990

In the early 1990s we, in the UK, had not even begun to attune our-selves to a changing climate, and in particular to the sharp decline in the frequency of extreme cold and heavy snow which manifested itself during the rest of that decade. The winters of the late 1970s and 1980s had delivered frequent episodes of severe weather – indeed, the mean Central England Temperature for the decade 1978 to 1987 was lower than for any other ten-year period during the twentieth century – and although the winters 1987–88 to 1989–90 were all mild and largely snow-free, the general feeling was that it was only a matter of time before the snows returned. Thus the winter of 1990–91 which had two very cold spells was regarded at the time as a reversion to normal. Subsequently, seriously heavy and widespread snowstorms have become appreciably less frequent, although our inability to cope with any sort of snowfall seems to have grown in parallel.

One slightly surprising aspect of the December 1990 snowstorm was the inadequate nature of the forecasts even 24 hours ahead. November had been mostly settled and dry with high pressure often in charge over the British Isles, and this anticyclonic weather lasted almost throughout the first week of December. The computer pre-dictions indicated a withdrawal of the high pressure system to mid-Atlantic on Friday the 7th and Saturday the 8th, allowing a frontal system to swing southwards across the UK, but most radio, tele-vision and newspaper forecasts spoke about a period of rain lasting perhaps three to six hours, with snow largely confined to high ground. Only the *Today* newspaper, now sadly defunct, indicated any appreciable accumulation of snow in the heavily populated areas of northern England and the Midlands, but even their offering considerably underplayed the event.

A close analysis of the synoptic developments during the days before the snowstorm explains why the forecasts were so poor. The cold front which tracked south-eastwards across the country during the second half of the 6th and the first half of the 7th separated air which had stagnated over the British Isles for several days from a polar-maritime airstream which had travelled across a broad expanse of relatively warm Atlantic Ocean from its source region in the Canadian Arctic. Neither of these air masses was particularly cold. During the first week of December, temperatures over Britain had typically been between 1°C and 4°C at night, although the night of the 5th/6th was frosty over much of England and Wales and between 5°C and 8°C by day with occasional local excursions as high as 10 or 11°C. The polar airstream had been much modified during its journey across the Atlantic, and the temperature on its arrival in northern Scotland at 6pm on the 6th ranged from 4°C at Lerwick in Shetland, Kirkwall in Orkney, and Stornoway on the Isle of Lewis, to 7°C at Tiree in the Inner Hebrides. At this stage, snow was confined to high ground above 400 metres in the north-west highlands. There was, however, a further change of air mass, from the polar–maritime of Canadian origin to an arctic–maritime one, originating within a few hundred kilometres of the north pole and streaming directly from the north. But this truly cold air lay some distance behind the main cold front, and was expected to arrive several hours after the front had cleared away from southern England. The weekend forecast therefore indicated a mixed bag of snow showers and sunny intervals on both the 8th and the 9th, the snow showers affecting mainly coastal districts exposed to the north wind.

What the computer models failed adequately to pick up until the evening of the 7th was a slowing in the south-eastward progress of the first cold front. This was a response to the persistently south-westerly flow in the upper atmosphere above southern Britain; this flow was oriented along the length of the front rather than across it, so the push from the north-west simply ran out of steam. This had two immediate effects. As the front stalled over northern England and Wales on the morning of the 7th, a shallow wave-depression formed on it, quickly developing a circulation of its own. At the same time, the advancing Arctic air mass rapidly caught up with the original front, and when this air entered the circulation of the new depression, the now much larger temperature contrast across the frontal zone injected additional energy into the system, causing the low-pressure centre to intensify abruptly. When the depression was first identified over Cumbria at 6am on the 7th, its central pressure was 1,006 millibars; by noon it lay near Leeds at 1003 millibars, by 6pm near Sheffield at 997 millibars, by midnight on the 8th over Lincolnshire at 992 millibars, by 6am near Cambridge at 988 millibars, and by noon over Essex at 987 millibars. The very cold Arctic air plunged rapidly southwards on the left-hand flank of the deep-

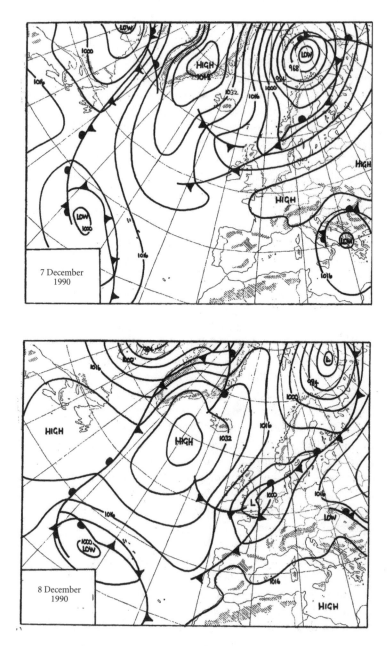

Figure 4.5 Synoptic charts for 7–10 December 1990

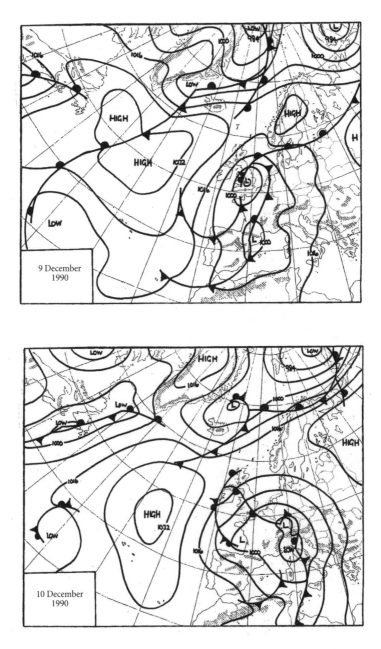

Figure 4.5 continued

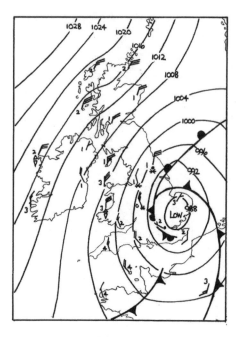

Figure 4.6 Detailed weather chart for the British Isles, 12 noon on 8 December 1990

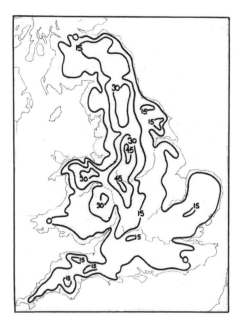

Figure 4.7 Maximum level snow depth (in centimetres), 8–9 December 1990

ening depression, and as it deepened the pressure gradient between it and the Atlantic high-pressure system steepened sharply, resulting in a marked strengthening of the northerly wind which now reached force 5 to 7 inland, and up to force 9 or 10 on exposed sections of the coasts of north-east England, Wales and Cornwall. The depression was at its deepest during the early afternoon of the 8th, after which it began very slowly to fill, though by now it was practically stationary over south-east England. On the 9th, still slowly filling, it began to move first south-westwards to Brittany, and thence southwards across the Bay of Biscay and into northern Spain; it was responsible for disruptive snowfalls in many parts of France and Spain between the 9th and 11th.

As soon as the Arctic air flow caught up with the original cold front, rain was replaced by wet snow as the temperature dropped several degrees. This was first observed on the evening of Friday the 7th over the higher ground of southern Scotland and northern England, quickly extending overnight to low ground in northern England and also spreading southwards into Wales and the Midlands, and on the morning of the 8th (though relatively briefly) to much of southern England between Devon and Kent, and to East Anglia. The Midlands was the worst hit region with heavy wet snow, driven by a near-gale northerly wind, falling for 15–18 hours without a break – throughout daylight hours on Saturday the 8th.

Table 4.2 Six-hourly weather reports at Birmingham airport, 7–9 December 1990

	00h 7th	06h 7th	12h 7th	18h 7th	00h 8th	06h 8th	12h 8th	18h 8th	00h 9th	06h 9th	12h 9th	18h 9th
Wind direction	SW	SW	SW	NW	NW	NNW	NNW	NNW	N	ENE	ESE	NNE
Wind speed	10	10	15	10	5	20	20	5	5	5	10	10
Barometer	1021	1015	1006	999	995	991	991	991	993	997	1005	1007
Temperature	6	6	6	3	3	0	0	1	1	1	2	2
Weather	fine	rain	cloud	rain	rain	heavy snow	heavy snow	cloud	sleet	cloud	fine	rain

Critical during this snowstorm was the temperature, and because temperature falls in a moist atmosphere at a rate of 1°C for every 150 metres elevation, height above sea level was crucially important, too, in assessing the character of the snow and the degree of disruption caused. Broadly speaking, while it was snowing, the temperature was around 1°C at sea level, 0°C at 150 metres above sea level, and –1°C at 300 metres above sea level. This meant that over the lowest-lying ground – around the Humber, for instance, and as far inland as York, Doncaster and Scunthorpe, also in the Fen district – the snow fell wetly and struggled to accumulate on the ground, and it was therefore reasonably easy for the highways authorities to keep the roads clear. Up to an altitude of about 150 metres (it varied slightly both geographically and with time, so these are broad-brush assessments) the snow had a high water content which discouraged drifting but at the same time made the snow very sticky and clinging, while above 150 metres the water content decreased, the snow became drier and more powdery in character, and it drifted much more readily. Drifting, then, only became a serious problem in the Peak District, the hills of the Welsh Marches (the Long Mynd, the Clee Hills, Wenlock Edge and the Malverns), over the Cotswolds, over the Clent and Lickey Hills in the Birmingham–Black Country conurbation, and over the wolds of east Leicestershire, Rutland, Lincolnshire and the East Riding of Yorkshire, and it was in these areas that the snow was deepest – 40 to 50cm deep in upland parts of Derbyshire and in the hillier parts of Birmingham and the Black Country. However, over the greater part of the Midlands where the temperature remained fractionally above zero the character of the snow was heavy and sticky, but it still accumulated to a depth of 20cm or more, and this led to paralysis on road and rail, to the disruption of power and phone services as cables snapped under the weight of ice and snow, to the failure of water-pumping stations resulting from extended power outages, and to the extensive damage to trees, shrubs and wildlife.

Large sections of the Midlands motorway network ground to a halt, and by Saturday afternoon the authorities estimated that, for a time, stationary traffic occupied over almost 200 kilometres of the M1, M5, M6, M40, M42, M45 and M54. The worst stretch occupied some 50 kilometres of the northbound carriageway of the M6 in Warwickshire, between the M1 junction and Birmingham, and here

Plate 4.3 A scene on the M1 motorway, just north of Nottingham, showing three lanes of southbound traffic at a halt, while northbound lanes appear abandoned, during the snowstorm of 8 December 1990. Large parts of the Midlands motorway network were gridlocked for over 24 hours following the sudden and unexpected onset of heavy snow and high winds

(Empics/PA Photos)

some motorists were stuck in their vehicles for 48 hours before they could be rescued. The police blamed the thoughtlessness of other motorists who had abandoned their vehicles on the motorway; they were only able to reach cars in the middle of the queue by physically removing those which had been left unattended from the ends of the traffic jam one by one. Shropshire's County Surveyor – in charge of snow clearance on local roads – responded pithily to accusations of incompetence:

> The chaos was caused by too many drivers continuing to travel in deteriorating conditions. They had a touching faith in their own driving ability and an expectation that roads would never be slippery, even when snow was falling. Drivers abandoned vehicles which blocked roads for the snow ploughs. One plough got stuck in some snow, but could not even reverse because vehicles had followed too closely behind. People demanded to be dug out of a drift so that they could finish their Christmas shopping. On Sunday, scores of motorists took whole families out to take pictures of the snow, and then complained when they got stuck. The M54 was kept clear until blocked by an accident caused by drivers overtaking the snow plough.

Questions were asked in the House of Commons, and the Transport Secretary of the day, Malcolm Rifkind, promised a review of arrangements for dealing with severe weather. When the review finally appeared, some six months later, the Department of Transport had decided that the combination of meteorological and other circumstances on 8–9 December was so exceptional that, however road-clearing services were organized, they would have failed to cope with this particular snowstorm. The almost unprecedented set of factors was outlined by the present writer in a contribution to the advice given to the Secretary of State:

> The worst hit regions – the Midlands, east Wales, Yorkshire and County Durham – had one to two hours of heavy rain before the snow set in, washing away salt and grit laid by the various highways authorities following warnings provided by weather forecasters. This is the nightmare scenario for these

authorities: there is no optimum time for gritting the roads under these conditions, for if they wait until the rain gives way to snow it is already too late. Treating the roads effectively prevents falling snow of light to moderate intensity from settling, but a treatment laid down after the snow has started to accumulate has little impact on what is already on the ground.

Over much of the Midlands, the snow began to fall heavily just before dawn on a Saturday morning when motorway traffic levels are at one of their lowest points of the entire week; steady traffic helps to keep treated roads clear of snow.

The snow had a very high water content, making it wet, soft, sticky, clinging and heavy. Water acts as a lubricant, and under some circumstances a covering of slushy snow can be more slippery than ice. Powdery snow drifts readily when the wind strength reaches 15mph, but wet snow does not normally drift unless sustained wind speeds exceed 30mph. This happens but rarely in inland districts except over exposed high ground, but in the Birmingham area, for instance, the sustained wind exceeded 35mph for a time on Saturday morning. Gusts of 50 to 60mph were recorded elsewhere in the region.

The snow fell at a temperature close to or a fraction above zero over much of the Midlands, hence its wetness, but once it had stuck to telephone and power lines, vehicle windscreens, trees and hedgerows, it was subject to evaporative cooling by the wind, turning the slush to ice. This had a terminal effect on many overhead cables and British Rail pantographs.

The 'plastering' effect due to wind and wetness resulted in sharply diminished visibility in motor vehicles which could have led to some drivers slowing or stopping with little or no warning. This was probably exacerbated by piles of accumulated slush and ice flying off heavy goods vehicles at frequent intervals during the day.

The rate of snowfall is known to be highest at temperatures close to zero, and on Saturday the rate of accumulation exceeded six centimetres per hour at times – very unusual in lowland Britain, and exceptionally rare over such a wide geographical area and for such an extended period of time. Once traffic stopped, for whatever reason, fresh snow meant that within minutes it was incapable of restarting.

Given this exceptional combination of factors, it is quite conceivable that the centralized emergency response body proposed by several critics would have failed to do any better. In this respect it is worth pointing out that the conjunction of meteorological conditions outlined above would be even rarer in other European countries – often held up as examples – where snow is more commonly dry and powdery, rates of accumulation are lower, and accompanying winds hardly ever reach the strength experienced over England on 8 December. In short, the Department would be unwise to establish a new response mechanism which would probably fail to cope with a repeat of the December 1990 snowstorm.

Having said all that, one clear improvement can easily be made. Information about severe weather as presently communicated to the general public on radio and television seems to be misunderstood or ignored by large sections of the population. Severe weather warnings, carefully and precisely worded, to be used only on occasions of truly exceptional weather, should be broadcast. These would have greatest impact if presented by continuity announcers or newsreaders, and kept clearly separate from routine weather forecasts.

In the wake of the storm the snow did not last very long. The temperature climbed a few degrees above zero in desultory sunshine on the Sunday afternoon, initiating a slow thaw, and the thaw accelerated on the Monday and Tuesday. Nearly all of it had gone by the morning of Thursday the 13th.

Chapter 5

Fog and smog: two case studies

What is smog?

Dr Harold Des Voeux is little known these days, although he was, in the first part of the twentieth century, a prominent figure in the fight to limit the pollution of our atmosphere. But he ought still to be remembered for one thing: he invented the word 'smog'.

The dangerous mix of dense water-droplet fog and thick smoke belching from industrial and domestic chimneys which characterized London's winters in Victorian times continued during the first half of the twentieth century. It was in his capacity as the treasurer of the Coal Smoke Abatement Society and member of the Committee for the Investigation of Atmospheric Pollution that, in 1905, Dr Des Voeux coined the word to describe that health-threatening concoction of smoke and fog. It sounded so singularly appropriate that it caught on at once.

'Smog' is used rather differently in the twenty-first century. These days it usually describes heavily polluted summer days and is more properly called photochemical pollution. This happens when strong sunlight reacts with the various nitrogen oxides – products of industrial processes and motor-car emissions – to create substantial concentrations of ozone in the lowest hundred metres of the atmosphere. This terminology first saw the light of day in the large conurbations of the USA where the motor-car and sunshine first joined forces to create this unpleasant form of pollution way back in the 1930s. It was first recognized in the UK in the long, hot summer of 1976, but increasing levels of exhaust pollutants on our increasingly congested roads have resulted in a regular recurrence since then during any sustained spell of sunny weather with light winds.

Ozone in the lowest part of the troposphere, where we live, is a toxic substance; this is in contrast to the ozone layer in the upper atmosphere where it plays a vital role in filtering harmful ultra-violet radiation coming from the sun. Low-level ozone causes irritation to the eyes, nose and throat, and for particularly vulnerable people such as the very young and very old, and those with heart or lung complaints, it can lead to serious and prolonged discomfort.

Until the mid-1970s the word was only used, at least in the United Kingdom, in Des Voeux's original sense. Those smoke-fogs remained very much a feature of the British climate, especially in and around our big industrial conurbations, until the various Clean Air Acts of the 1950s and early 1960s. But since then they have almost completely vanished, along with such descriptive and whimsical terms as 'pea-soupers', 'great stinking fogs' and 'London particulars'.

Air pollution was well known in ancient Rome where buildings were blackened with soot, and the health of the citizens was occasionally compromised by the dense pall of smoke which settled over the imperial capital in cold, calm weather during the winter months. Indeed, until the rapid urbanization which began with the Industrial Revolution just over 200 years ago, the compactness of cities and the very high population density within them resulted in heavy concentrations of pollutants. This was certainly true of most British cities; residential property lay cheek-by-jowl with smithies, lime kilns, potteries, bakeries and other small-scale industry.

The smoke problem increased markedly when coal replaced wood as the dominant fuel during the 1200s. According to Peter Brimblecombe, who wrote *The Big Smoke* – the definitive history of air pollution in London – the first recorded complaint about thick smoke pollution in the UK was made by Queen Eleanor in the year 1257, while the first attempt to address the problem was put forward in 1285 when a commission was established to investigate London's rapidly deteriorating air quality. Its recommendations, however, were largely ignored.

Over the next six centuries or so, legislation came in dribs and drabs and had little worthwhile effect. As the capital grew, so the atmospheric pollution became progressively worse. Nor was London alone: the great nineteenth-century conurbations of the Midlands and northern England, the Greater Glasgow area, and the industrial sprawl across the coalfields of south Wales, all suffered chronic smoke pollution especially during the winter half-year. Periodically, when the winter air became stagnant and cold, fog and smoke conspired to create serious 'smog' episodes.

London fogs were a constant preoccupation of those Victorian writers who lived and worked in the capital. Several, including Arthur Conan Doyle and Robert Barr, produced apocalyptic visions of London strangled by a poisonous pall of pollution with millions of inhabitants perishing as a consequence. Others, Dickens and Byron among them, used descriptions of old-fashioned smoke-fogs, typical nineteenth-century 'pea-soupers', to conjure images of contemporary London.

E. F. Benson was another who wrote freely, roughly a century ago, about the metropolis choking on its own smoke, but one paragraph of his that particularly catches the attention described the sudden and dramatic clearance of a typical 'London particular':

> From a sick dead yellow the colour changed to gray, and for a few moments the street seemed lit by a dawn of April; then across the pearly tints came a sunbeam, lighting them with sudden opalescence. Then smoke from the house opposite, which had been ascending slowly like a tired man climbing stairs, was plucked away by a breeze, and in two minutes the whole street was a blaze of primrose-coloured sunshine.

This is a splendid metaphor for the longer-term change in the winter climate of Britain's conurbations – not just London – since the passage through Parliament of the Clean Air Acts in 1956 and 1968.

Over half a century has now passed since the great London smog of early December 1952, the inevitable disaster which finally precipitated legislation with sufficient teeth to rid London of its smoke for good. Between 6,000 and 7,000 people, most of them already suffering from chronic heart and/or lung diseases, died prematurely. Most published accounts quote a figure of 4,000 deaths, but this was the total for the old London County Council area, and there were at least 2,000 more victims in the urbanized parts of Essex, Kent, Surrey, Middlesex and Hertfordshire. Some studies estimate that the true death toll may have been as high as 12,000 or even more, as the death rate remained above the long-term average for several weeks beyond the two-week period which comprised the original calculation.

Even after this appalling human disaster, the Conservative government of the time showed considerable reluctance to take serious legislative action, and it was left to one of their backbenchers, the flamboyantly mustachioed Gerald Nabarro, to introduce a private member's bill. This finally provoked the government into action, and the first Clean Air Act at last became law in July 1956.

Although since the 1950s we have had serious problems in stagnant anticyclonic weather with high concentrations of nitrogen dioxide, carbon monoxide and (in summer) ozone, urban smoke in the UK is now a thing of the past. As one would expect, fog frequency and density in our conurbations has diminished markedly. And the main consequence of this has been a marked increase in winter sunshine.

Detailed sunshine measurements have been made in the United Kingdom since 1876 when daily records were first kept at the two London observatories at Greenwich and Kew. An improved sunshine recording instrument was introduced at several of our weather stations in 1881, and the number of sunshine-measuring sites around the UK increased sharply during the last twenty years of the nineteenth century. Meteorologists at the time were deeply interested in the problem of pollution in our towns and cities, and several sites were established in urban areas which rapidly illustrated how serious the problem was. A few of these sunshine-recording stations were introduced specifically to extend our knowledge of city-centre sunshine, notably at Bunhill Row in London and Oldham Road in Manchester.

The information culled from this network of stations speaks for itself. The last standard reference period before the Clean Air legislation covered the years 1921–50, and some of the records obtained for the Greater London area were as shown in Table 5.1.

Several of the sites have changed between the two periods, but the evidence is unequivocal. Before 1950, central London lost around three-quarters of its sunshine to the pall of pollution during December and two-thirds of it during January; November and February both lost about a half. Since 1960 there is no discernible geographical pattern to the distribution of monthly sunshine hours.

Although London is Britain's largest urban area, it was not during the earlier part of the twentieth century its dirtiest or smokiest conurbation. December sunshine aggregates before 1950 in the

Greater Manchester area included just seven hours at Oldham Road, thirteen hours at Whitworth Park, thirteen hours also at Burnley, and fifteen hours at Bolton, contrasting with 35 hours in rural Cheshire. Across the Pennines the average was nineteen hours at Bradford, 21 hours at Huddersfield and 23 hours at Wakefield, compared with 40 hours at Harrogate and York. In Glasgow, Springburn Park's twenty hours contrasted with 32 hours on the Firth of Clyde.

Table 5.1 Monthly sunshine records for London, 1921–90

LONDON	*Distance from centre (kilometres)*	*Monthly sunshine in hours*					
		Oct.	*Nov.*	*Dec.*	*Jan.*	*Feb.*	*Mar.*
1921–50							
Bunhill Row	0	80	31	11	17	37	87
Kingsway	0	82	34	15	22	39	89
Regent's Park	2	83	41	27	29	45	92
Hampstead	7	100	51	37	40	59	112
Kew	14	95	50	38	42	59	111
Croydon	18	103	52	42	46	63	117
Wisley	42	96	52	41	47	62	117
Farnborough	54	108	63	46	53	70	125
1961–90							
High Holborn	0	111	69	49	48	63	112
Hampstead	7	110	67	50	48	65	112
Kew	14	110	69	47	52	71	114
Heathrow	24	108	68	46	52	68	110
Wisley	42	107	64	43	48	65	111
Gatwick	44	110	64	46	51	68	117

Note: some of the stations have incomplete records

With the dramatic increase in motor-vehicle emissions over the last 40 years we cannot say that Britain's air is necessarily that much cleaner than it was before the Clean Air Acts, but it is certainly much clearer. Our industrial areas enjoy more winter sunshine now than they have done since the eighteenth century, and it is not far-fetched to suggest that the City of London now has sunnier Decembers and Januarys than at any time in the last 800 to 900 years.

The 'great stinking fog' of December 1873

Mention London smogs, and most people will think of 'that one back in the 1950s', not necessarily because they experienced it but rather because there have been so many fiftieth-anniversary accounts of it – written and broadcast – during the early years of the present decade. The great London smog of December 1952 was certainly not unique, even in the twentieth century, but it did give the final impetus to the drive for clean air legislation, first in the metropolis and then in other conurbations around the country; the first Clean Air Act eventually passed through Parliament in 1956.

The frequency of fog and smog in London had appeared to be in decline since about 1900, although it is difficult to be certain because estimates and measurements of horizontal visibility were only made systematically in the UK from the First World War onwards, in response to the demand for such information by wartime aviators. Before that, weather observations contained notes of 'fog', 'mist' and 'haze', but no specific visibility thresholds were laid down, and such reports were inevitably subjective, varying from observer to observer, and probably even varying through time with the same observer.

What is not in doubt, however, is the fact that piecemeal legislation during the early part of the twentieth century had reduced the pollution loading of London fogs with a significant decline in the quantity of both particulate matter (in simple terms, soot) and sulphur dioxide between the 1890s and 1950s. In recent times, sophisticated computer modelling of the relationship between the concentrations of different pollutants on the one hand, and the density and frequency of fogs on the other, suggests that the available fog statistics for the period 1850–1900 in London are, despite the limitation noted above, entirely consistent with the known levels of pollution.

Frederick Brodie, the Meteorological Office's chief climatologist in the late nineteenth century, drew together all available fog observations in the capital, chiefly using daily reports from Westminster and Brixton, noting an average of 60–65 foggy mornings per year during the period 1871–90, with the months October, November, December and January being the worst affected. There were 86 such mornings in 1886, 83 in 1887, and 75 in both 1873 and 1889. Between 1900 and 1950 the average was between 45 and 50, and since the Clean Air Act came into force in 1956, between 10 and 15. It is hardly surprising, then, that in late Victorian London heavily polluted fogs were so frequent that they were routinely the topic of conversation, and a subject often addressed by writers, from journalists in the morning papers to humorists in magazines like *Punch*, and to poets and novelists, Dickens included. The nineteenth-century writer and poet, Thomas Miller, described the typical London fog of the 1850s:

The whole city seems covered with a crust, and all the light you can see beneath it appears as if struggling through the huge yellow basin it overspreads. You fancy that all the smoke which had ascended for years from the thousands of London chimneys had fallen down all at once, after having rotted somewhere above the clouds; smelling as if it had been kept too long, and making you wheeze and sneeze as if all the colds in the world were rushing into your head for warmth, and did not care a straw about killing a few thousands of people, so long as they could but lodge comfortably for a few hours any where.

Another late-Victorian meteorologist who took a special interest in London fog was the Honourable Francis Albert Rollo Russell, the third son of the Liberal Prime Minister Lord John Russell, and the uncle of philosopher Bertrand Russell. Rollo Russell, as he was generally known, was a frequent contributor to the *Quarterly Journal* of the (later, Royal) Meteorological Society and *Symons's Meteorological Magazine*, and he explained the differences, as then understood, between London fogs and fogs in rural areas surrounding the capital:

A London fog is a complex phenomenon, greatly differing from a country fog, and since it does not depend altogether on the natural action of temperature, winds and vapour, may be dealt with as a removable evil. The following are the chief distinctions between a London and a country fog. A country fog is white, without smell (unless perhaps a slight odour of ozone), and not disagreeable to breathe. It seldom thickens after the first hour after sunrise. It is pure condensed vapour, and therefore clean. The sun appears perfectly white when seen through it. A London fog is brown, reddish-yellow, or greenish, darkens more than a white fog, has a smoky, or sulphurous smell, is often somewhat dryer than a country fog, and produces, when thick, a choking sensation. Instead of diminishing while the sun rises higher, it often increases in density, and some of the most lowering London fogs occur about midday or late in the afternoon. Sometimes the brown masses rise and interpose a thick curtain at a considerable elevation between earth and sky. A white cloth spread out on the ground rapidly turns dirty, and particles of soot attach themselves to every exposed object.

This was the background against which the 'great stinking fog' of December 1873 occurred, a fog that was described at the time by Dr Robert James Mann, the President of the Meteorological Society, as 'the worst for many decades, perhaps since 1814'. The fog closed in across the whole of London on Tuesday 9 December, and finally lifted again on Saturday 13 December.

The autumn of 1873 had already given England a taste of the quiet, stagnant foggy weather that was to come. High pressure settled over the country between 27 and 30 October, bringing thick fog to many places during the nights and mornings, and although in most places the sun burned it off quickly, in the capital fog lingered for three days. Fog returned on 4 and 5 November, and again on the 9th. High pressure returned between the 13th and 21st, during which extended period the weather was characterized by 'high fog' – what we would today call a layer of stratus cloud, perhaps 300 to 600 metres aloft, below which was trapped much smoke and other pollutants. The last week of November was unsettled with frequent rain and vigorous westerly winds, clearing away the pollution very effectively.

Another large anticyclone drifted slowly eastwards across northern France and southern England between 1 and 5 December, but once again the sky over London remained largely overcast, preventing the formation of fog. A cold front travelled south-eastwards across the UK later on the 5th, and in its wake skies cleared as pressure again rose strongly across the British Isles. Fog formed widely in south-east England and East Anglia early on Tuesday the 9th, and although it cleared for a time during the day even in the metropolitan area, it clamped down quickly as the sun set while the temperature dropped well below the freezing point. At Brixton that night a minimum value of −7°C was recorded, and it did not again rise above freezing in some outlying parts of the Home Counties until midday on the 15th. The high-pressure system was anchored over southern Britain until late on the 14th when it finally collapsed ahead of an active Atlantic frontal system. Thick fog persisted until the 14th, not just in the capital but across large parts of southern England, East Anglia and the Midlands, although in London it was loaded with progressively higher concentrations of soot, sulphur dioxide and other industrial effluent. Most of the remainder of December was changeable and largely smog-free, although transitory ridges of high pressure allowed fog to form temporarily in London on Christmas Day and again on the 28th and 29th.

We may turn to Rollo Russell again to add some detail concerning the character of the fog, and some personal observations of its effect on the capital city and its inhabitants.

The fog of the middle of December 1873 was one of the thickest and most persistent of this century, and deserves to be noticed in some detail; not because it differed from the ordinary London fog, except in intensity, but because the character and effects of these fogs are unusually conspicuous in this instance. I was residing at Richmond Park at the time, but, having occasion to go daily to London, thought it worth while to pass through several districts of the town for the purpose of making observations on this remarkable fog. After about a week of cloudy, misty, and quiet weather, on the night of December 8–9, a hard frost came on, with a scarcely perceptible air from the west. The 9th was a very fine day in the country and at the western suburbs, but in London there was a dense black fog all day, and many accidents occurred on the river and in the streets. On December 10 the frost continued.

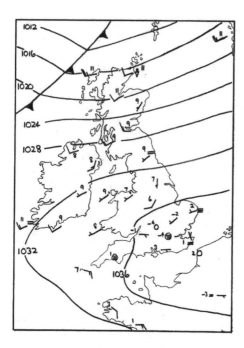

Figure 5.1a Detailed weather chart for the British Isles, 8am on 9 December 1873

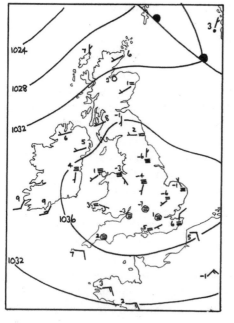

Figure 5.1b Detailed weather chart for the British Isles, 8am on 12 December 1873

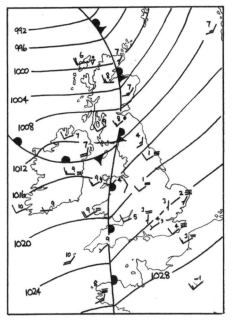

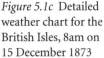

Figure 5.1c Detailed weather chart for the British Isles, 8am on 15 December 1873

At 8am the thermometer was 22°F. The calm continued, only a very light air from west prevailing near the ground, but higher up from north-east. The smoke of London was drifting over Richmond from the north-east. The barometer remained very high. In London, the weather was the same as that of the day before, and many of the fat cattle exhibited at the Great Show at Islington died of suffocation. It was not possible during a great part of the day to see across a narrow street, and in the evening a choking sensation was felt in breathing. On the 11th, the barometer stood at 30.62 inches of mercury, the thermometer at 22°F. At Richmond the weather was fine, with mist, a dead calm, then a lower current from the west and south-west, and cirro-cumulus clouds from east-north-east. Foggy in London. On the 12th, barometer steady, thermometer about 32°F; a thick and damp ground-haze. Not so bad in London. Dead calm, but local currents as follows: Vauxhall, south-east; Ludgate Hill, south-east; Camden Town, north-west; Richmond, west. Cirro-cumulus were still moving from the east. On the 13th, a ground-haze at Richmond, cloudy, and dead calm. No very thick fog in west London. Barometer still about 30.64in. On the 14th, very slight drizzle in London about 9am. Dead calm. Current north-west at Belgrave Square; south-east in the City and at Shoreditch at 10am. Dark yellow fog, rapidly thickening northwards from the City, till at Haggerston pitch-black darkness at 10.15am, the lower air being tolerably clear. About a mile of this darkness like that of night then gradually lighter towards the west, and at Camden Town fair, at Hampstead pretty clear and cloudy, clouds hanging at an elevation of about 350 feet above the sea. At all stations north and west of Haggerston (on the North London line) a north-west current. In the afternoon a gentle breeze from west-north-west. Thus ended this long period of calm. The deaths exceeded the weekly average by about 700.

Russell's note on the number of deaths attributable to this smog event is often seen quoted in accounts of the great London smog of 1952 when 7,000 or more perished, perhaps in an attempt to indicate how much worse the 1952 event was than anything else for which we have records. Dr Mann tells us in his Presidential Address on 21 January that the death rate in the metropolitan area – almost exactly coincident with the later Administrative County of London which did not come into existence as a county council until 1889 – was 23 per 1,000 in the week preceding the fog, 27 per 1,000 in the week of the fog, and 38 per 1,000 in the week following, and that there were 2,396 deaths from respiratory ailments during the three-week period, which was some 850 above the normal number. But these statistics do not include any casualties noted in subsequent weeks when the mortality rate remained elevated, nor do they include deaths which occurred in that part of the London urban area which extended into neighbouring health districts, including those parts of Middlesex, Surrey and Essex which were outside the metropolitan area boundary, and in particular the boroughs of Croydon, West Ham and East Ham. These last two, in the East End of London, were very densely populated and characterized by extreme poverty and deprivation, and people suffered poor health. When all these areas are included and the analysis extended for five weeks after the fog event actually occurred – in other words up to the middle of January – the aggregate death toll in the Greater

London area may be conservatively estimated to have been 2,800, and possibly substantially more than that.

Although most casualties were, as one would expect, the result of chronic illness being exacerbated by the high pollution levels, there was also significant loss of life through other causes. Dozens of people were reported missing, and some 20–25 bodies were later recovered from the London docks, and from canals elsewhere in the capital, presumably the result of accidental drownings, perhaps under the influence of excessive alcohol consumption. Pickpockets and footpads also enjoyed a profitable week, and one may surmise that some of their victims could well have been despatched under cover of the fog. In fact, the surge in the crime rate during the December 1873 smog excited a good deal of comment in the newspapers, and *Punch* magazine turned its attention to it as well, with these allegedly humorous suggestions to the Metropolitan Police:

First – Should the fog be very dense, withdraw half the Police from the thoroughfares. Remember their lives are valuable to the community at large.

Secondly – Let none of the Street Lamps be lighted, until the usual time (if then); they are of very little use, and the shops *must* have more blaze than usual. Never do for yourself what you can get some one else to do for you.

Thirdly – In the neighbourhood of St. Paul's and the Banks, where the traffic, like the Fog, is at its thickest, let care be taken to secure the absence of all light and all Police. Surely everyone who is out on such a day ought to be old enough and wise enough to take care of himself. As to omnibuses, waggons, carts, cabs and carriages, they ought all to have lamps, and, when they haven't lights, they have lungs, and can ward off danger by continuous shouting.

Fourthly – No extra Gas must be used at Railway stations, and great care should be taken that all the carriages may be left without the usual lamps. When the Fog has entirely cleared off, the Lamps may be lighted, and the Police may resume their duties.

During the latter half of the nineteenth century there was a growing social awareness of the deleterious effects of London's heavily polluted fogs, although such awareness was widely regarded as a fad of the chattering classes at the time (and indeed until the 1950s). Nevertheless, some writers attempted to calculate the material cost of the phenomenon, assigning figures to the water and soap used in personal bathing 'over and above what would normally be necessary'; the washing or cleaning of clothes, bedding, curtains and carpets; the increased frequency of repair and replacement of clothing, household fabrics and wallpaper; restoring the paintwork, gilding and metalwork to shop fronts, street boards, statues and monuments, and public buildings; the frequent requirement for window cleaning; the cost of additional candles, lamps, gas and other fuels, and increased frequency of chimney sweeping; and the more rapid depreciation in the value of houses and other property caused by dirt and decay. One such study estimated the cost, at 1873 prices, to be between £1.5 and £3 million per year. At today's prices that would be approaching 100 times as much; even so, it seems to be an extremely conservative estimate.

'Christmas can't getaway': the four-day fog of December 2006

The 'great fog' of 2006, as it will doubtless come to be called, has already spawned its own clutch of myths: an unprecedented mete-orological event, four days of continuous fog at Heathrow airport,

Plate 5.1 A British Airways Boeing 737 taxis out of Heathrow's Terminal 1 into the fog for takeoff, 21 December 2006
(Tim Ockenden/PA Wire)

Plate 5.2 Hundreds of passengers queue in Heathrow's Terminal 1
following British Airways' cancellation of 180 flights, 21 December 2006
(Tim Ockenden/PA Wire)

on a par with the great smogs of the nineteenth and twentieth centuries, British fogs are getting worse because of climate change, and
so on.

The facts are these. No one died in this particular fog from respiratory or cardiac failure triggered by the polluted atmosphere.
The total fog-related death toll in London was in single figures, the
consequence of a handful of road traffic accidents. However, the
number of heart attacks triggered by the stress of waiting, frustrated
and uninformed, in crowded airports, is not known. The network of
pollution monitoring stations across the capital noted slightly
elevated levels of particulate matter (carbon particles), but sulphur
dioxide, nitrogen oxides and carbon monoxide all remained resolutely in the 'low' category. This was a clean London fog, something
our forebears would scarcely have believed possible.

However, it is certainly a long time since we had such an extended
foggy episode over such a large part of England and Wales, but that
is entirely a reflection of the remarkable decline in the frequency of
fog in this country during the past 25 years or so. This followed an

earlier decline in the late 1950s and 1960s following the introduction of clean air legislation 50 years ago. Reasons for the more recent decrease in fogginess in Britain are difficult to pinpoint; they probably include a lower frequency of high-pressure systems over the UK and adjacent parts of Europe during late autumn and early winter, and a reduction in domestic and industrial pollution on the near Continent, as well as a continued improvement in air quality here in the UK. One has to go back to the early 1990s for the last examples of widespread and persistent fog in England – February 1993, January 1992 and December 1991 spring to mind.

Meteorological developments in advance of, during and after the foggy episode of December 2006 were very interesting, and closely matched the pressure patterns which prevailed during nineteenth- and early twentieth-century smogs. It hardly requires a detailed scientific analysis to deduce that, had the weather situation of December 2006 occurred a century earlier, with all the particulate and sulphur-based pollution of the time, it would have lasted twice as long as the great smog of December 1952. 'What-ifs' are regarded by some historians as sterile exercises, but it is salutary to think that, in an earlier time, this would certainly have been one of London's great killer smogs, and maybe the greatest of all.

Much of November and the first half of December 2006 had been very disturbed with frequent rain, often heavy and widespread, and an unrelenting wind blowing chiefly from the south-west. A marked cold front crossed the UK from the north-west on the 15th and 16th, followed by a substantial rise of pressure over the country, but a further shallow depression travelled south-eastwards across Ireland and south-western Britain on the morning of Monday the 18th before high pressure really asserted itself. By 6pm that day, though, the barometer had risen above 1030 millibars over practically the entire country, with a centre of 1034 millibars over southern Scotland. The anticyclone remained resolutely anchored over the British Isles throughout the next five days, reaching a peak intensity late on the 21st when a reading (adjusted to sea level) of 1046.3 millibars was observed at Sennybridge in the upper Usk valley, just north of the Brecon Beacons. This was the highest barometer reading anywhere in England and Wales since January 1992, and the highest in December since 1980. On Saturday the 23rd there was a subtle change, the centre of highest pressure shifting into

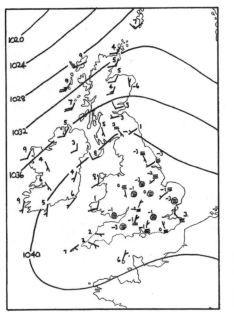

Figure 5.2 Detailed weather chart for the British Isles, 6am on 21 December 2006

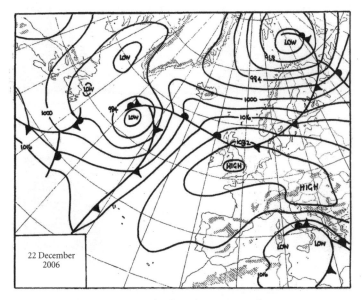

Figure 5.3 Synoptic chart for 22 December 2006

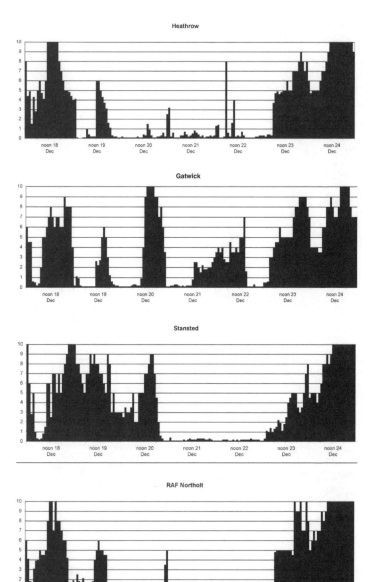

Figure 5.4 Visibility (in kilometres) reported from London's airports and airfields, 18–24 December 2006

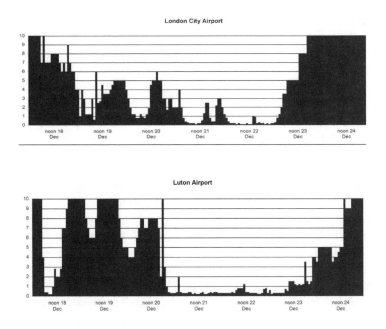

Figure 5.4 continued

the southern North Sea, and this allowed a gentle easterly or north-
easterly airflow to develop across southern and, later, central England, but the anticyclone remained centred just to the east of
England, now declining very gradually, until Wednesday the 27th.
On that day a cloudy south-westerly air flow finally dislodged the air
which had stagnated over the UK for very nearly ten days.

The air in the developing anticyclone on the 18th had originated
in a sub-polar air mass over the Atlantic Ocean between latitudes 50
and 60 degrees north, so it was rather mild and moist to start with.
There was additional moisture provided by the waterlogged countryside following the long wet spell, and in particular the rain that
fell in association with that day's shallow depression. Fog had
formed the previous night beyond the depression's influence in
eastern England from Norfolk to Yorkshire. Then during the afternoon and evening of the 18th, under cloudless skies, with very
nearly the minimum length of daylight, the winter solstice being
just a few days away, the temperature quickly fell to the dew point as

dusk approached, resulting in the formation of widespread fog over most of England, with patches also in east and north Wales, Northern Ireland, and the central belt of Scotland. The fog was dense in places, with reported visibility below 50 metres locally, but it was not deep, and even the modest ranges of hills of southern and central England – the Cotswolds, the Chilterns, the North Downs – stood out above the fog and enjoyed several days of cloudless skies and uninterrupted sunshine. But in the Thames Valley, lower-lying parts of the Midlands and Lincolnshire, and in the Vale of York, the fog which formed on the 17th or 18th persisted almost unbroken until the evening of the 22nd. The light east to north-easterly breeze which developed late on the 22nd resulted in the fog lifting into a layer of low stratus cloud, typically between 200 and 500 metres above sea level, which therefore shrouded all high ground in a thick hill fog while visibility at lower altitudes slowly improved.

In London's built-up area the fog was not quite as persistent as in some outlying districts around the Home Counties: the warmth generated by the city and slowly emitted by the bricks, concrete and asphalt lifted the temperature, at least in inner London, a degree or so above the dew point from time to time, and without the high concentrations of pollutants of earlier fog events in the capital this allowed visibility to improve a little here.

Thus the notion that Heathrow was continuously 'fogbound' for four days – widely touted at the time – requires examination. Of the 96 hourly weather observations from the 19th to the 22nd inclusive, nineteen were outside official fog limits with visibility of 1 kilometre or

08:30	Glasgow	BA1474	Cancelled
08:40	Aberdeen	QF3355	Cancelled
08:40	Glasgow	BD002	Flight closing
08:40	Edinburgh	BD052	Cancelled
08:45	Manchester	MH9658	Cancelled
08:55	Manchester	AA6621	Cancelled
08:55	Edinburgh	AA6562	Cancelled
08:55	Belfast City	BD082	Go to Gate
08:55	Dublin	BD123	Flight closing
09:05	Dublin	EI153	Go to Gate
09:15	Aberdeen	BD674	Delayed to 09:40
09:45	Newcastle	AA6720	Cancelled
09:55	Glasgow	AA6436	Cancelled
09:55	Durham Tees	BD332	Cancelled
09:55	Cork	EI711	Gates 80-90
10:05	Dublin	EI155	Gates 80-90
10:10	Shannon	EI373	Gates 80-90
		AA6564	Cancelled

Plate 5.3 British Airways announced it was cancelling all domestic flights into and out of Heathrow, 22 December 2006 (Tim Ockenden/PA Wire)

Plate 5.4 The layer of fog was relatively shallow, and ranges of hills such as the Downs and the Chilterns stood above it, enjoying blue skies and unbroken sunshine. Looking west from Dunstable Downs, Ivinghoe Beacon stands like an island in a sea of fog (compare Plate 8.3)
(Philip Eden)

more, only 26 were below the threshold for thick fog of 200 metres, and none was in the dense fog category with visibility below 40 metres.

It is true that the airport's Air Traffic Control was required by law to apply Low Visibility Procedures throughout that period, triggering the reduction of aircraft movements. But the huge number of passengers suffering long delays tells us how close to capacity Heathrow airport is, especially at holiday time, rather than whether the fog was unprecedented or not. In the event, British Airways were forced to cancel all domestic flights into and out of Heathrow for three days running, and (counting all airlines) there were 370 cancelled flights on the 20th, 284 on the 21st, and 74 on the 22nd from that single airport – an aggregate of 728. Dozens of flights were also cancelled from other British airports, notably Birmingham, Manchester, East Midlands, Norwich, Southend and Cardiff, bringing the national total to over 1,000. In all, approaching 200,000 travellers were affected.

Already some organizations choose to measure the severity of windstorms by insurance losses, and others assess the intensity of a drought by the water industry's inability to guarantee water supplies. That is up to them. But scientists should guard against redefining meteorological events in terms of the impact they have on human activity.

Chapter 6

Summer floods: two case studies

Rainfall measurements: intensity versus duration

Rainfall has three dimensions, each of which contributes to the characteristics of the flooding which follows a major downpour. These dimensions are intensity, duration and geographical extent. They do not, you may have noticed, include the traditional and still-ubiquitous measure of rainfall – quantity. That, and a variety of other parameters, can be calculated from the main three. The quantity of rain that falls at a given place is equal to intensity multiplied by duration; the quantity of rain that falls on a river catchment is intensity times duration times geographical extent – either of the catchment, or the rainfall event, or a combination of the two. And so on. And yet these three dimensions are rarely distinguished in discussions of floods found in newspapers or the broadcast media, and they are even sometimes glossed over in official reports.

As they are so infrequently examined in the non-specialist literature, most people are unfamiliar with the range of measurements that each of these parameters may exhibit – what is routine, what is unusual, and what is likely to trigger flooding. This was brought home during the prolonged downpours of June and July 2007. The grim reports of storm and flood were usually introduced by a newsreader telling us that one place or another had received 'a month's worth of rain in 24 hours', accompanied by a look of astonishment and a sharp intake of breath. Sometimes a reporter on site, togged up in wellies and oilskins and paddling in the safest bit of floodwater he or she could find, would be questioned about this particular statistic by the studio presenter, and would nod sagely and agree in reverential tones that it was hardly surprising that such a rare, once-in-a-lifetime event had produced such a severe flood.

What these journalists were actually doing was demonstrating their lack of knowledge, and they therefore failed to place the event into a proper statistical and historical context. 'A month's worth of rain in 24 hours' can be a useful phrase to describe 50mm of rain in a day – certainly a noteworthy downpour – but I wonder how many of those reporters (and therefore their viewers and listeners) realized that there had been eleven days during the first five months of 2007 with falls of 50mm or more somewhere or other in the UK. That is equivalent to one every two weeks on average. Here is some more context: during the Boscastle storm in August 2004 some 200mm of rain fell in one afternoon, that is four months' worth in four hours; and in a single storm just south of Middlesbrough in August 2003 almost 50mm of rain fell in little more than ten minutes, and that is a month's worth while the one o'clock news was on.

Let us examine in more detail what different kinds of rain amount to in numbers. First of all we may note that rainfall has traditionally been measured in linear units – originally in inches and latterly in millimetres. Although this is so ingrained in our national psyche that it seems natural, it is in fact entirely illogical: what we should actually be measuring is the volume of water. Fortunately, using metric units, one millimetre of rain is exactly the same as one litre per square metre of water, so the units are interchangeable; many countries still prefer to use millimetres, but l/m² may be encountered in, for instance, Spain and Italy.

Official weather observers have three categories of rainfall intensity: light rain falls at an average rate of 0.5mm per hour or less, moderate rain at between 0.5 and 4mm per hour, and heavy rain at more than 4mm per hour. We can perhaps relate better to these rates by applying some everyday descriptions, remembering that a strengthening wind or falling temperature can make the rain even more disagreeable:

0.1mm/hr: A gentle drizzle that most people will venture out in without an umbrella; Wimbledon brings on the covers but Test cricketers continue to play; roads and footpaths don't get wet; motorists find 'intermittent wipe' for their windscreens to be too frequent.

0.2mm/hr: A steady drizzle or spitting rain; many people raise umbrellas, while those without raincoats may hurry a little even though they are only getting slightly damp; damp patches also on roads; motorists use 'intermittent wipe'.

0.5mm/hr: A fast drizzle or light rain; umbrellas in general use, anyone in shirtsleeves will get wet, albeit quite slowly; roads and footpaths become very damp and puddles form slowly; motorists can't quite decide whether to use the 'intermittent' or the 'normal' setting for windscreen wipers.

1.0mm/hr: Steady rain (if there is a strong wind, driving rain); pedestrians in raincoats and hats begin to hurry, those in shirt-sleeves look for cover; footpaths wet with puddles, road gutters begin to run; 'normal' setting for windscreen wipers is now definitely appropriate.

4.0mm/hr: A steady downpour, raining cats and dogs; rain bounces off pavements so that even those with large golfing umbrellas will get wet below the knee; pedestrians in coats run for cover; large puddles form quickly; motorists use 'fast wipe' and should also have headlights on and reduce speed.

10mm/hr: A heavy downpour, raining stair-rods or pitch-forks; most people seek cover until the worst of the rain has passed; motorists find that 'fast wipe' can still cope, but should use headlights and cut their speed sharply to avoid aquaplaning; floods appear where drains are choked.

25mm/hr: A torrential downpour, will lead to extensive local flooding if it lasts more than, say, half an hour; 'fast wipe' can barely cope and some motorists may choose to pull over.

100mm/hr: A cloudburst; such a rate of rainfall rarely lasts more than five minutes in the UK; all motorists should pull over as horizontal visibility may drop to less than 50 metres; sloping roads turn into torrents; drains and sewers may fail; eccentric ark-builders get busy.

We also need to think about the effect that the duration of the rainfall event will have. A passing shower may deposit 0.1mm in five minutes, scarcely wetting the ground; a sharp April shower may drop 1mm in ten minutes and send everyone racing for shelter; while one of those 'freak' hailstorms which, far from being freaks, are so typical of Britain's less settled summers, can leave 10–20mm in the rain gauge in, say, twenty minutes. Severe thunderstorms typically provide 25 to 50mm in an hour or two, while one of those long-lasting summer downpours which occurred so frequently in summer 2007 may give us 100mm in 48 hours. Winter rains are normally much heavier over the windward slopes of upland Britain – for example, Snowdonia, the Lake District and the Western Highlands of Scotland – and in a well-established moist south-westerly air flow in November or December as much as 400–500mm may fall in these districts over the course of five days or a week. All of these events, from the twenty-minute summer hailstorm upwards, can trigger flooding, but the character of the flooding depends on the way different river catchments (and in towns and cities, the drains and storm sewers) respond to the downpour.

From the earliest days of rainfall recording and analysis the importance of rainfall intensity, duration and geographical extent were recognized, but instruments to measure these elements simply did not exist. By contrast, it was relatively easy to measure rainfall amounts daily or weekly; in its most basic form a rain gauge is little more than a bucket with a scale, although it was immediately evident that standardization of materials, size and exposure would be necessary if results from different places were going to be compared one with another, and it was also clear that a method of measuring very small amounts of rain would be essential. George James Symons, the father of the British rainfall network, developed all these (and many other) aspects of rainfall reporting in the 1850s and 1860s. Autographic rain gauges were designed during the later decades of the nineteenth century to provide for the continuous measurement of rainfall on a chart attached to a clockwork drum, and thus were able to measure rainfall intensity at given sites. But these instruments were complicated and very expensive, and it was financially impracticable to introduce them at more than a handful of sites around the country. For observers who had to persevere with the traditional manual rain gauge, Symons asked them to be scrupulously attentive during heavy rainfall events and to note down the start and finish times of the downpour and to make regular measurements during the storm if they could. Many of the observers, being what we might now call 'anoraks', were delighted to oblige. Symons did not appear to like maps very much, but his successors as Directors of the British Rainfall Organization, Sowerby Wallis, Hugh Robert Mill and Carle Salter, constructed detailed maps of major rainfall events, so from their time onwards we have excellent qualitative and quantitative information about each of the three important rainfall parameters.

As the quantity of statistical data accumulated – and accumulate it did, at a very fast rate – it became possible to identify the frequency with which different types of downpour occurred in the UK, and to deduce what severity of flooding might ensue. We should remember here that researchers in the early twentieth century had no computers, were working with manuscript ledgers, had to undertake all the calculations themselves, and were dealing with hourly data from over 300 autographic gauges as well as daily figures from more than 6,000 observing sites – something like 5 million pieces of data per year. That sort of never-ending data-handling exercise in the pre-computer age was a truly Herculean labour.

As early as 1888 Symons devised a scheme which subjectively divided short-period heavy rainfall events into four categories which he described as 'Exceptional', 'Remarkable', 'Worthy of note', and 'So common as to be unworthy of special tabulation'. There were occasional relatively minor adjustments to the divisions and to their descriptions during the next few decades, and the calculations were put on a much more rigorous mathematical footing by the well-known climatologist Ernest Bilham in 1935, but essentially the same scheme for examining heavy rainfall events continued until the early 1960s. The three working categories became 'Very rare', 'Remarkable', and 'Noteworthy'. Further adjustments were made by introducing contemporary statistical techniques by D. J. Holland in 1961, but the method had fallen into disuse a decade later. From the mid-1960s, the application of computer technology to the problem meant that it became relatively easy to calculate return periods – that is, the frequency with which a particular fall of rain in a particular time period might be expected to recur on average – and then from the early 1980s the development of the rainfall radar over the

UK (and other countries) introduced an entirely new range of techniques for assessing both the statistical probability of flooding over a long period and also the development of individual flooding events in real time.

We can get some idea of the limits of rainfall intensity over given periods by analysing the data over the past 120 years or so from the entire autographic rain-gauge network – a couple of dozen of them during the late nineteenth century, but they numbered in the hundreds from the 1920s onwards. But our knowledge of these limits must remain imperfect because the nature of extreme rainfall events is that they affect extremely localized areas, and an array of a few hundred gauges scattered across the country will clearly be unable to sample these downpours adequately. It is therefore extremely probable that the most exceptional of them will have fallen through the gaps in the recording network. Indeed, experienced climatologists suggest that the real limit of rainfall intensity for a given time-period may approach twice the observed extreme.

We can also get a feeling for near-instantaneous rainfall rates from the latest generation of rain gauges which are found in fully

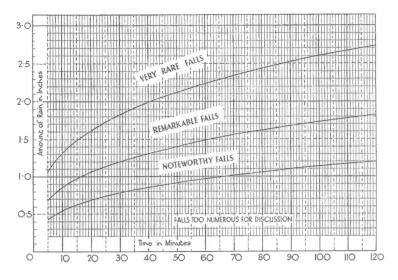

Figure 6.1 The 'Bilham Diagram', categorizing heavy falls in short periods

automated electronic weather-recording stations, but which were actually introduced in the early 1970s in a stand-alone form by some water companies and river authorities. These are known as 'tipping-bucket' gauges, and operate by collecting the rain alternately in two attached containers which pivot about a central axis as first one and then the other fills with water. As each bucket tips, an electrical pulse is transmitted to the data-logger, and the water pours away. The most common 'bucket' size is 0.2mm, so at a rainfall intensity of, say, 60mm per hour it takes twelve seconds to collect 0.2mm, at 300mm per hour it takes 2.4 seconds, and at 600mm per hour just 1.2 seconds. As it takes between half a second and a second to complete the tip from one bucket to the other, it can be seen that this type of rain gauge, although it works very effectively at most rainfall rates, rapidly becomes inefficient during rainfall of an exceptional intensity. Given that caveat, it appears that a one-second rainfall rate close to or slightly more than 1,000mm per hour may represent the limit for the UK, although it should be added that the equivalent rainfall rate during an extreme hailstorm will undoubtedly be greater. (As a comparative exercise, I placed a standard rain gauge under the splendid power-shower in my bathroom and collected 213mm of 'rain' in one minute – a rate of 12,780mm per hour, which is about twelve times as strong as its most violent natural equivalent!)

For finite periods of time we may mention several remarkable downpours which have occurred over the years. At Carlton-in-Cleveland, some 15km south of Middlesbrough, 50mm of rain fell in a single short-lived thunderstorm on the morning of 10 August 2003; the date will be remembered more widely for the record-breaking heat which afflicted south-east England where the temperature climbed to 38.1°C at Kew Gardens in south-west London. These two extreme events were not unconnected as the intensity of individual rainstorms is, in part, dependent on ambient temperature and humidity levels. During this storm the bulk of the rain, 45.9mm, fell in just ten minutes. The UK's largest rainfall amounts measured for given time-spans are as shown in Table 6.1.

We can plot these observations and calculate a line of best fit, and it is gratifying to observe that all but one of them fit very closely indeed. If we then accept the experts' opinion and double the figures that we read off the graph, we get the results shown in Table 6.2,

Table 6.1 Largest UK rainfall amounts

Hours	Minutes	Amount (mm)	Location	Date
	2	16.0	Carlton-in-Cleveland, N. Yorkshire	10 August 2003
	5	30.0	Carlton-in-Cleveland, N. Yorkshire	10 August 2003
	8	41.3	Carlton-in-Cleveland, N. Yorkshire	10 August 2003
	10	45.9	Carlton-in-Cleveland, N. Yorkshire	10 August 2003
	12	50.8	Wisbech, Cambridgeshire	27 June 1970
	15	55.9	Bolton Hall, Lancashire	18 July 1964
	20	63.5	Sidcup, Greater London	5 September 1958
	20	63.5	Hindolveston, Norfolk	11 July 1959
	25	67.0	Pershore, Worcestershire	11 June 1970
	30	80.0	Eskdalemuir, Dumfries	26 June 1953
	45	97.0	Orra Beg, Co. Antrim	1 August 1980
1	00	97.0	Orra Beg, Co. Antrim	1 August 1980
1	15	101.6	Wisley, Surrey	16 July 1947
1	30	117.0	Footholme, Lancashire	9 August 1967
1	45	152.4	Hewenden Reservoir, West Yorkshire	11 June 1956
2	00	193.2	Walshaw Dean Lodge, West Yorkshire	19 May 1989
3	00	193.2	Walshaw Dean Lodge, West Yorkshire	19 May 1989
4	00	197.0	Otterham, Cornwall	16 August 2004
5	00	215.9	Cannington, Somerset	18 August 1924
6	00	220.0	Cannington, Somerset	18 August 1924
12	00	279.4	Martinstown, Dorset	18 July 1955

which we may therefore describe as representing the likely upper limit for rainfall amounts in the British Isles for given time periods:

Table 6.2 Likely upper limits for rainfall amounts

Minutes	Amount (mm)	Hours	Amount (mm)
1	20	1	220
2	31	1.5	265
3	41	2	300
4	50	3	340
5	58	4	370
10	90	5	395
15	112	6	415
20	129	9	465
25	143	12	500
30	155	18	575
45	190	24	620

The Norfolk floods, August 1912

Mention Norwich and canaries in the same breath and most people will think of the city's football club, nicknamed The Canaries. Few outside Norfolk know that Norwich was a centre of canary breeding for a long time, although in recent decades the interest in caged birds has waned. Incidentally, before moving to its present head-quarters in Carrow Road the football club's ground was known as The Nest. There are social and demographic reasons why canary breeding has been in decline since the early twentieth century, but the trigger may have been a single meteorological event.

In late August 1912, Norwich was visited by a flood unprecedented in the city's annals. Vast tracts of Norfolk lay under water, especially around the eastern fringes of Norwich's built-up area where the Rivers Yare and Wensum converge, and also downstream

Plate 6.1 The great Norfolk flood of 26–27 August 1912. A view down
Magdalen Street, Norwich
(Eastern Daily Press)

Plate 6.2 Midland Street, Norwich, during the great Norfolk flood
of 26–27 August 1912
(Eastern Daily Press)

towards the Broads where the environmental consequences were severe and long lasting. Norwich was cut off by rail and road for over two days, the local tram service was suspended, and the supply of both electricity and town gas failed. Over 40 bridges were washed away, thousands of hectares of crops were destroyed, and about 4 square kilometres of housing in the city were inundated, the depth of water reaching 4 metres in some places.

Given the severity and extent of the flood, it was a surprise that no more than three people were drowned. Canary breeding was very much a spare-time activity of the working men of the district, and the birds were typically kept in cages in sheds in the gardens and backyards of the lower-lying part of Norwich. This was the first area to be flooded, and the speed of events and the imminent threat to human life meant that little attempt was made to save the birds, thousands of which therefore perished. Local press reports indicated that one particular strain of Norwich canary was completely wiped out.

The August 1912 flood in Norfolk was the result of a prolonged downpour lasting about 30 hours, from the early hours of the 26th

Plate 6.3 Boatmen rescuing residents of Midland Street, Norwich, in the
great Norfolk flood of 26–27 August 1912
(Eastern Daily Press)

Plate 6.4 The remnants of Horstead Bridge across the River Bure, broken
down by the power of the floodwaters, about 12km north-east of
Norwich, following the great Norfolk flood of 26–27 August 1912
(Eastern Daily Press)

until mid-morning on the 27th, and it came at the end of the coldest, wettest and dullest August on record. The depression responsible for the downpour formed in mid-Atlantic on the 24th in an area of slack pressure in the wake of the preceding disturbance. It then travelled eastwards along the fiftieth parallel, to be just north of Brittany overnight on the 25th/26th, by which time it was entraining some very warm and humid air from the region of Madeira; the contrast between this plume of warm air and the relatively cold air elsewhere in its circulation provided the energy for the depression to deepen, which it did as it travelled east-north-eastwards along the English Channel, thence turning north-north-east into the southern North Sea. The original analysis of the event which was carried out by the Director of the British Rainfall Organization, Dr Mill, describes its progress thus:

> At 7am on August 26th a small but well developed depression had its centre off the east of Kent, and during that day moved northwards very slowly, being situated north-east of Cromer at 6pm. On the morning of August 27th the centre was in the east of the North Sea, the track having apparently turned rather abruptly to the east during the evening or night of the 26th. We have pointed out on previous occasions that abrupt changes in the direction of travel of the centres of low pressure systems have been associated with exceptionally heavy rainfall.

This was very perspicacious of Mill, for he was writing before meteorologists had much systematic knowledge of the influence of circulation patterns in the upper air, and the way that depressions on sea-level weather charts are actually steered by the airflow aloft. With this in mind, we may now deduce the upper-air pattern, and by also examining very closely the available weather observations made at the time we may re-analyse developments as follows.

A well-developed low-pressure system in the upper atmosphere travelled steadily east-north-eastwards across Wales, the Midlands, Lincolnshire, thence across the North Sea between 25 and 27 August. Initially, the depression on the sea-level chart lay to the south of the centre aloft, and was steered in an east-north-easterly direction along the English Channel. But by the time it reached the Straits of Dover the surface 'low' had run some distance ahead of the

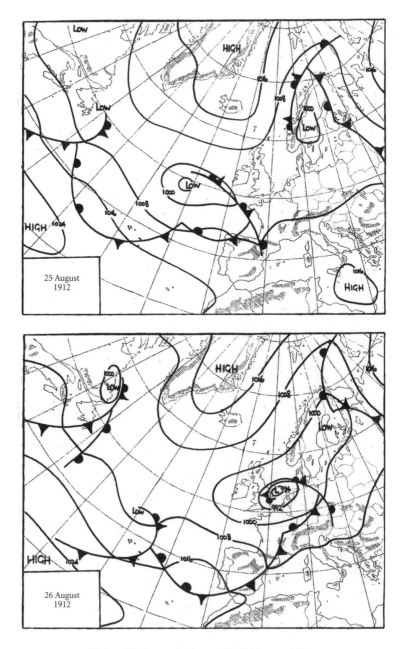

Figure 6.2 Synoptic charts, 25–27 August 1912

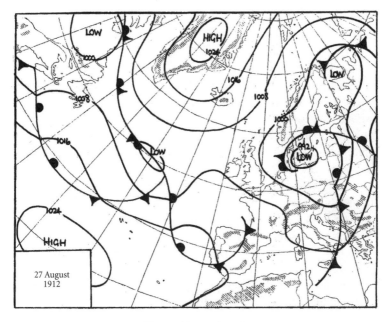

Figure 6.2 continued

upper-air 'low' and as a consequence the steering flow aloft backed south-westerly and then southerly, sending the depression north-wards, now decelerating, past the Thames Estuary and along the coast of East Anglia. By this stage, the upper 'low' caught up again, passing directly above the surface 'low', causing it to come almost to a complete halt, perhaps even heading a little to the north-west to a position offshore Cromer, and at the same time deepening substan-tially to become a very active disturbance with a central pressure of 978 millibars, exceptionally low for August. As the upper 'low' finally moved ahead of its surface equivalent, the westerly steering flow on its southern flank now kicked in, and the surface depression acceler-ated away eastwards towards northern Holland. The protracted loitering of the depression off north-east Norfolk during its most vigorous phase was associated with the long period of very heavy rain focused on that north-eastern segment of Norfolk. It is instruc-tive to observe that this deceleration and deepening followed by a change of direction – all resulting from the interaction of the surface and upper-air depressions – appear time and time again in analyses of extreme weather in the UK. For example, both of the snowstorms featured in Chapter 4 displayed this sequence of events.

A total of 205.5mm fell at Brundall which is located 8km to the east of Norwich city centre – approximately four months' worth of rain in just over one day. Rainfall for the event exceeded 150mm across 1,300 square kilometres of countryside, and the aggregate rainfall averaged across the entire county of Norfolk was 125mm. This is the equivalent of almost 150,000 million gallons or 650 million tons of water. Put another way, it is as if the entire contents of Lake Windermere had been emptied twice over.

The heaviest falls were all found in the eastern half of Norfolk, and included a measurement of 210.1mm at Hanworth Hall, just south of Cromer, and one of 209.5mm at Sprowston Council School on the north-eastern outskirts of Norwich, but both of these figures were discounted in Mill's analysis owing to unspecified instrumental problems. However, neither is so different from neighbouring reports as to be unambiguously wrong. Of those observations accepted by Mill, the heaviest falls were:

> 205.5mm at Brundall
> 185.2mm at Worstead (The Grange)
> 193.5mm at Old Catton
> 185.0mm at Norwich (Carrow House)
> 190.7mm at Norwich (Heigham Cemetery)
> 184.1mm at Reepham
> 190.5mm at Sprowston (Old Lodge)
> 183.4mm at Dunston Hall
> 186.9mm at Norwich (Ipswich Road)
> 182.6mm at Worstead (Westwick Gardens)
> 186.7mm at Rockland St Mary
> 182.6mm at Sprowston Hall
> 186.4mm at Norwich (Eaton)
> 181.1mm at Moulton
> 186.2mm at Heathersett
> 178.1mm at Swainsthorpe

There were no recording rain gauges in Norfolk in 1912, but Mr John Willis, the observer at the Grange, Worstead, checked the rainfall every hour between 9am and 6pm on the 26th. He noted a

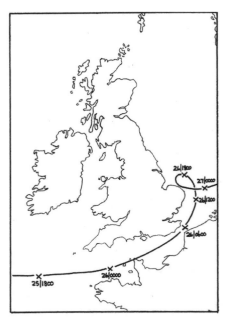

Figure 6.3 Track of depression, 25–27 August 1912

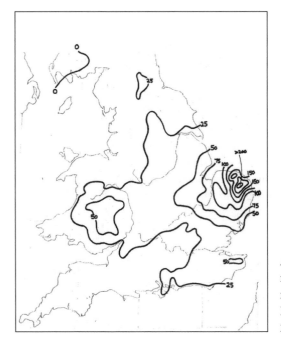

Figure 6.4 48-hour rainfall (in millimetres) over England and Wales, 26–27 August 1912

gradual increase in intensity to a maximum during the early after-noon; between midday and 1pm he measured 29.2mm and between 1pm and 2pm 27.9mm, but no other hour exceeded 20mm. Arthur Preston of Eaton, Norwich, described the day:

> It was an awful visitation and I shall never forget it. It was not a straight, hard, thunder-rain, but reminded me more of a blizzard of fine snow . . . The floods are terrible. The train service, tram service and electric light service all discontinued, and many of the streets and roads in low-lying parts of the city are like rivers, and in the cathedral close there is a large lake. There is no record of anything like it in East Anglia, the 'great Norwich flood' of November 1878 sinking into insignificance beside it.

In fact, there was a record of historic floods inscribed on a wall in the city, and the 1912 event exceeded the previous record, set way back in 1614, by almost 40cm. There is, however, anecdotal evidence of an even greater flood in some of the villages to the north-east of the city in 1607.

As the depression began to move away and the rain eased, a ferocious westerly gale sprang up, uprooting thousands of trees, demolishing walls weakened by the floodwaters, and rendering rescue work increasingly difficult. According to the local papers, one boatman, William Marrison, stayed in his boat for twelve hours on the 27th, ferrying over 100 people from their flooded houses to dry land. Local authorities were quick to act, opening school buildings in the higher parts of the city for the thousands of homeless, and a relief fund raised just over £20,000 (£1.6 million at 2008 prices) to help the victims. There were great fears for the safety of the people of Great Yarmouth and Lowestoft, with the floodwaters of the Yare, Bure and Wensum rivers converging just downstream of Norwich, but the Norfolk Broads and the peaty soils surrounding them acted as a giant sponge and saved the two coastal towns from serious inundation. However, the great bulk of sediment carried down Norfolk's main rivers was deposited in the Broads, wrecking its fragile ecosystem, and silting up the lakes and water-courses which in turn had serious repercussions for water-borne commerce, the burgeoning leisure-craft activity, and the boat building industry in the district.

The summer floods of 2007

In September 1954 the Editor of *Weather*, the Royal Meteorological Society's house magazine, wrote: 'At the end of a summer whose coolness and wetness have provided subject matter for cartoonists and ideas for advertisers, it is now possible to recall that there have been such summers before. Situated as we are at the downwind end of the Atlantic, this summer's weather it not so unlikely that atomic explosions, flying saucers, or condensation trails ... need be invoked in explanation.' The same sort of search for a culprit happened during the rain-sodden summer of 2007. However, summer 1954 was much the colder and gloomier of the two, and rain fell copiously throughout, though not with such devastating effect as happened in 2007.

The first signs that summer 2007 might be rather out of the ordinary appeared in early spring. Throughout eight long weeks there was little or no rain over large parts of the UK, and April was the warmest and one of the driest ever recorded. Extended spring droughts are often followed by poor summers, and there is a reason for it.

The main driving force behind the depressions and anticyclones which bring us our variable day-to-day weather is the jet stream, a conveyor belt of strong winds which encircles the northern hemisphere, normally found in middle latitudes, and located some 10 to 15 kilometres above the earth's surface. The Rocky Mountains straddle the hemisphere and act as an obstruction to the jet stream, so that, rather as an obstruction in a river causes a series of standing waves downstream, the jet stream follows a series of sinusoidal waves downwind of the Rockies. When the jet stream is at its most powerful, these waves are elongated and flat, and the winds in the upper atmosphere blow from the west, give or take a couple of points of the compass, right around the planet. When the jet stream is relatively weak, the waves are shorter and much more pronounced, so the airflow aloft meanders widely, first poleward, then equatorward, then poleward again. These giant troughs and ridges are responsible for the character of our seasons: when a ridge gets locked into a position over western Europe for several weeks or months, Britain's weather becomes persistently dry, warm and sunny, but if we are stuck under a trough, it just goes on raining.

The northern hemisphere's circulation has been in a predominantly weak phase since 2003 – some experts like to call this a negative phase of the Arctic Oscillation, and with respect to Europe you will also hear the expression 'a negative North Atlantic Oscillation index' – with the result that Britain's weather has lurched sharply from deluge to drought and back to deluge again. During the transitional seasons of spring and autumn, the atmosphere is in flux because of the seasonal increase or decrease in heat energy coming from the sun, and the upper-air ridges and troughs rarely get stuck in the same place for more than a few weeks. There is a well-known tendency for the wavelength to get shorter as summer approaches, with the result that, if a ridge has dominated western Europe's weather during the spring, it is very likely to shift westwards into mid-Atlantic by early summer; if it shifts far enough, then the British Isles will lie under the next trough in the sequence. During high summer the energy received from the sun changes little, so it is quite common for the arrangement of ridges and troughs, once established in late June, to persist for the rest of the summer – quite common, but not inevitable.

Thus previous extended spring droughts, especially those centred on April, have often given way to wet summers, as in, for instance, 1997, as Table 6.3 shows.

Table 6.3 Rainfall, spring and summer 1912–2007

Year	April rain (percentage of normal)	Summer (June, July, August) rain (percentage of normal)
1912	14	207
1954	24	156
1957	15	134
1974	21	123
1980	30	141
1982	36	127
1997	37	145
2007	17	175

During the last century the only exceptionally dry April which failed the test was in 1984, and that year May and early June were outstandingly wet, but the remainder of the summer was warm and dry. The opposite trend can be identified too in Table 6.4.

Table 6.4 Rainfall, spring and summer 1913–2006

Year	April rain (percentage of normal)	Summer rain (June, July, August) (percentage of normal)
1913	158	58
1935	164	90
1937	131	82
1959	126	78
1983	183	54
1989	138	79
2000	158	86
2006	140	76

The correlation between April rainfall and the character of the following summer is surprisingly strong – good enough to be used as a crude seasonal forecasting tool, but with enough exceptions to require caution in applying it.

The rains which set in during the second week of May 2007 continued with little intermission until the very end of July. The character of the season was dominated by the repeated occurrence of exceptionally heavy and prolonged cyclonic rains, as distinct from the more typical torrential but short-lived thundery downpours, although there were several of these as well. The cyclonic episodes occurred around 14 and 27 May, 14 and 24 June, and 20 July.

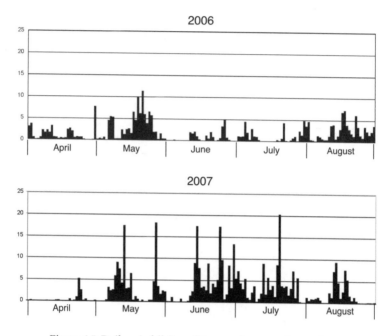

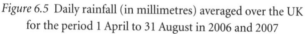

Figure 6.5 Daily rainfall (in millimetres) averaged over the UK
for the period 1 April to 31 August in 2006 and 2007

13–14 May 2007

The depression responsible for this event crossed the Atlantic close
to latitude 45°N, unusually far south, enabling it to draw into its circulation very warm and very moist air from south of the Azores and
south of Madeira. This was crucial in delivering the large quantities
of rain that fell over much of the country; in fact, averaged over
England and Wales, rainfall for the 13th alone was 20mm (24mm
for the entire event), equalling the record for a single day in May, a
record that was to stand for less than two weeks. As the depression
approached the Bay of Biscay it turned to the left and tracked across
the UK along a line from Cornwall to Lincolnshire; its progress was
slow, and the heaviest and longest-lasting rain lay on the northern
flank of the depression's track – over Wales, the west and north
Midlands, South and West Yorkshire, and Lincolnshire. Highest
individual rainfall totals included:

82mm at Preston Montford (Shropshire)
55mm at Emley Moor (West Yorkshire)
54mm at Telford (Shropshire)
52mm at Shawbury (Shropshire)
51mm at Tenbury Wells (Worcestershire)
48mm at Hawarden (Flintshire)
47mm at Chester (Cheshire)

After the long dry period during the earlier part of the spring, all rivers had been running very low until the 13th, so no major rivers overtopped their banks as a result of this 30-hour-long soaking. Nevertheless, there was extensive surface flooding in and around both Shrewsbury and Telford, and to a lesser extent in Wrexham and Chester. This surface flooding led to a large number of reports of road closures and traffic jams on the morning of the 14th in those four districts.

27–28 May 2007
The depression responsible for this event had originated over sub-tropical waters of the Atlantic Ocean near Bermuda, thence travelling first in a north-easterly direction, and then eastwards across the ocean close to latitude 50°N. The intensity of the rain was fuelled by the remnants of tropical air in the storm's circulation, but this time – as in 1912 – it was the interaction between the surface depression and the upper-air depression which extended the life of the downpour to between 40 and 48 hours in the worst hit areas. Averaged over England and Wales, 21mm fell on the 27th alone, and 30mm over the whole of the event, establishing a new record for May. Worst hit areas included the Isle of Wight, south Hampshire and south-west Sussex, the northern Home Counties, and much of East Anglia. The flood in Luton was the worst for at least 115 years in spite of recent flood control measures put in place along the River Lea; the Luton International Carnival, scheduled for the 28th which was a bank holiday, was abandoned at the last moment as much of the route for the carnival procession was under water, as indeed was the procession's destination in Wardown Park. Heaviest falls for the 27th were:

87mm at Luton (100.4mm in 49 hours starting at 8pm on
   the 26th)
75mm at Beccles (Suffolk)
75mm at St Catherine's Point (Isle of Wight)
72mm at Ventnor (Isle of Wight)
66mm at Royston (Hertfordshire)
64mm at Wantage (Oxford)
59mm at Hempstead-by-Holt (Norfolk)
59mm at Whipsnade (Bedfordshire) (69.7mm in 49 hours
   starting at 8pm on the 26th)
58mm at Calthorpe (Norfolk)

13–15 and 24–25 June 2007
June 2007 was certainly an exceptional month, and local rainfall
records were broken in the Birmingham, Nottingham, Lincoln, Hull
and Sheffield districts – not surprisingly, the areas which were worst
hit by flooding during the third and fourth weeks – but rainfall
totals varied widely from one part of the country to another.

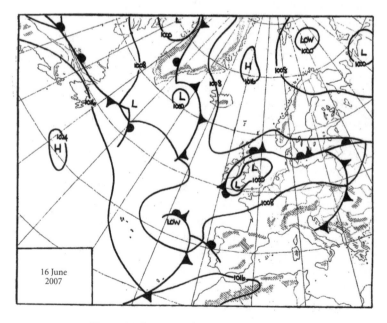

Figure 6.6 Synoptic chart for 16 June 2007

Northern Scotland had a very dry month, and in Orkney and Shetland it was one of the driest Junes on record, while many parts of southern England reported only small excesses. In Greater London, for instance, rainfall was just 25 to 35 per cent above normal and it was the wettest June in the capital only since 1999. Similar figures to London's were reported from the region covering Oxford, Bedford and Cambridge, and also from the Plymouth and Cardiff districts. Averaged over England and Wales it turned out to be the wettest June since 1860, just pipping 1997 which was the wettest of the twentieth century.

There were quite large variations within Yorkshire's boundaries, too, from 2.8 to 4.8 times the average (Table 6.5).

As with the mid-May event, the depression which brought the rain between 13 and 15 June travelled across the Atlantic at very low latitudes, entraining warm and very moist air from the sub-tropical

Table 6.5 Rainfall amounts, Yorkshire, June 2007

Location	13th–15th	23rd–25th	June total	Percentage of the long term average
	mm	mm	mm	
Leeming	103	21	178	413
Dishforth	92	41	165	393
Topcliffe	68	29	120	285
Linton-on-Ouse	70	64	161	393
Carlton-in-Cleveland			168	280
Loftus	68	32	142	273
Fylingdales	93	117	269	427
Bridlington	68	74	183	345
Wilsden	123	71	283	398
Emley Moor	115	85	272	484
Church Fenton	75	61	172	419
Leconfield	57	64	161	366
Howden	61	61	151	387

Atlantic, before turning to the left and tracking across the UK from the Bristol Channel to the Humber. Upwards of 50mm again fell over a large part of the west and north Midlands, and also in Northern Ireland, but it was Yorkshire that caught the brunt. Road flooding caused a great deal of disruption in the Birmingham and Nottingham areas, as well as in the conurbations of West and South Yorkshire, and some homes were affected too. Highest 72-hour rainfall totals at individual stations were:

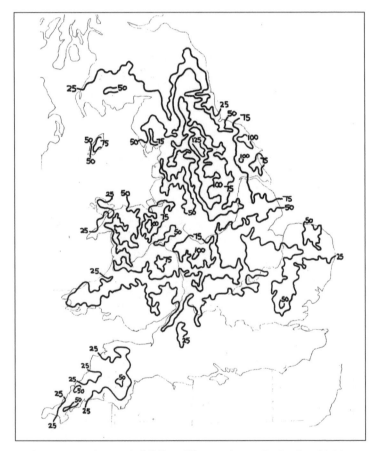

Figure 6.7 48-hour rainfall (in millimetres) over England and Wales,
14–15 June 2007

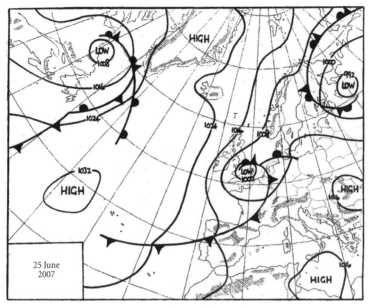

Figure 6.8 Synotic chart for 25 June 2007

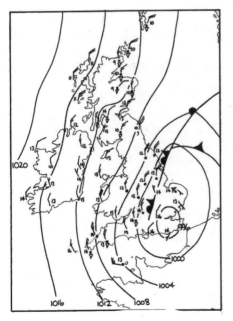

Figure 6.9 Detailed weather chart for the British Isles, 6pm on 25 June 2007

192mm at Trassey Slievenaman (Co. Down)
146mm at Lumley Moor (North Yorkshire)
142mm at Harlow Hill (North Yorkshire)
138mm at Birstwith Hall (North Yorkshire)
135mm at Weston Park (Sheffield)

The devastating floods which affected many parts of Yorkshire, but especially Sheffield, the Don valley and Hull, from the 25th onwards, were the result of a double whammy. It was hardly ever mentioned in press and media reports at the time, but most parts of Yorkshire actually had more rainfall and the area affected by heavy rain was much greater during the earlier downpour, between the 13th and 15th, when the resultant flooding was appreciably less severe and less widespread. But by the time the second downpour arrived, the water table was very high, the flood plains were sodden, river levels remained very high, and there was simply nowhere for the billions of gallons of additional water to go. More than 100mm of rain fell during the 48 hours of the 14th–15th over approximately 2,400 square kilometres, and during the 24th–25th over about 1,100

Plate 6.5 Motorists attempt to negotiate deep flood waters on the A63 dual carriageway, heading into Hull from the west, 25 June 2007
(Anna Gowthorpe/PA Wire)

Figure 6.10 48-hour rainfall (in millimetres) over England and Wales, 24–25 June 2007

square kilometres. This late-June flood left four people dead, 30,000 were evacuated from their homes in Hull, and a further 700 who lived in the shadow of Ulley reservoir, near Rotherham, and 40,000 residents of Sheffield, were without electricity for24 hours or more. The emergency services said that they had rescued 3,500 people from rising waters, and that there had been an additional 4,000 callouts. Highest 48-hour rainfall totals at individual stations for 24–25 June were:

121mm at Winestead (East Riding)
119mm at Keyingham (East Riding)
118mm at Great Culvert (Hull)
117mm at Fylingdales (North Yorks)
113mm at Hull
112mm at Tickton (East Riding)

19th June

We cannot leave June without a mention of the destructive flash flood in the middle Severn valley late on 19 June. Thunderstorms broke out widely that afternoon and evening, and a violent storm led to a destructive flash flood at Hampton Loade, just south of Bridgnorth, in Shropshire.

One striking aspect of June's weather was the fact that the mean monthly temperature ended up over a degree above the average for the standard reference period 1971–2000, indicating that the under-lying warming trend is identifiable even during our dullest and wettest months.

The reporting of the Yorkshire floods of 2007 also lacked a sense of perspective, and very rare was the piece-to-camera or newspaper column which attempted to place the several rainfall events of the summer in a proper statistical and historical context. We have no-toriously fallible memories when it comes to newsworthy weather events, but here in the UK our records of rainfall and temperature are second to none, providing us with a descriptive and statistical background to the British climate extending back to the late seven-teenth century. However, statistics can be easily misused. We should guard against those who argue from the particular to the general, suggesting that, for instance, because it was the wettest month on record at one site in Sheffield that we could then call it Britain's wettest month on record. If events are too easily labelled 'a new record' or 'unprecedented', then the people we pay to keep our infra-structure running have a ready-made excuse for failure, and those local councillors who allow developers to build on flood plains against expert advice are not held to account for their greed and incompetence: 'It's never happened before so how could we be expected to plan for it?'

June may have been the wettest month on record at Weston Park, Sheffield, but 25mm more rain fell in Leeds in June 1982, while a similar amount fell in Longdendale, west of Sheffield, in July 1973. On both occasions there was severe, widespread and long-lasting flooding in different parts of Yorkshire. The county is also particularly prone to short-lived flash flooding following violent thunderstorms: June 2005 in Helmsley, May 1989 in Halifax, the western suburbs of Bradford in August 1956, and Ilkley in July 1900 all spring to mind. So no one should be allowed to escape with that 'never happened before' line.

19–21 July 2007

A shallow depression developed over central France on the afternoon of the 19th on the boundary between a mass of exceptionally hot air covering much of southern and eastern Europe, and much cooler air over the British Isles and the Bay of Biscay. This contrast, together with a deepening area of low pressure in the upper atmosphere, favoured intensification of the surface depression as it travelled slowly north close to the Greenwich meridian. The centre of the 'low' lay over the eastern Channel at midnight on the 19th/20th, over Norfolk at midnight on the 20th/21st, and near Aberdeen at midnight on the 21st/22nd. Heavy rain fell over most of England, Wales and eastern Scotland, but the downpour lasted longest – between 30 and 36 hours – over the south-west Midlands and neighbouring parts of Wales, and this was where the largest rainfall totals (covering the two days 19th and 20th) were found:

163mm at Sudeley Lodge (Gloucestershire)
157mm at Pershore Agricultural College (Worcestershire)
140mm at Chastleton (Warwickshire)
140mm at Langley (Gloucestershire)
138mm at Winchcombe (Gloucestershire)
136mm at East Shefford (Berkshire)
133mm at Tewkesbury (Gloucestershire)
128mm at Brize Norton (Oxfordshire)
115mm at Pershore airfield (Worcestershire)
115mm at Little Rissington (Gloucestershire)
107mm at Malvern (Worcestershire)

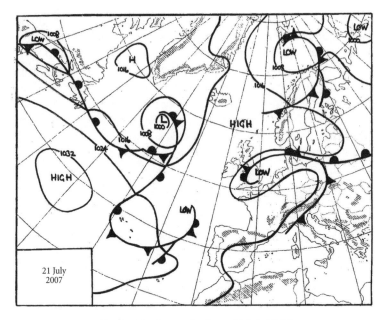

Figure 6.11 Synoptic chart, 21 July 2007

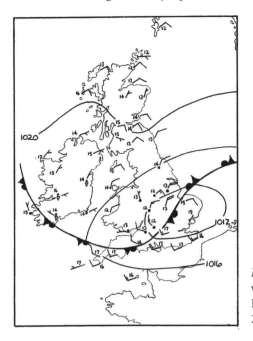

Figure 6.12 Detailed weather chart for the British Isles, 6pm on 20 July 2007

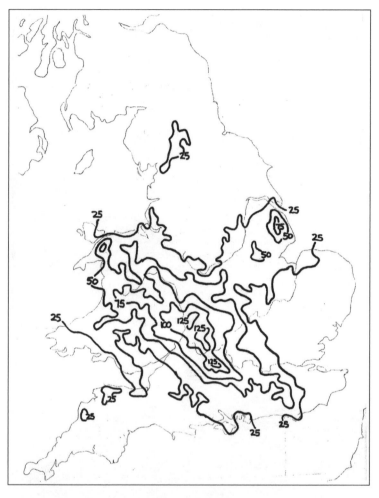

Figure 6.13 48-hour rainfall (in millimetres) over England and Wales,
19–20 July 2007

Plate 6.6 Aerial view of the flooded River Severn, just north of
Tewkesbury, Gloucestershire, on 22 July 2007
(Barry Batchelor/PA Wire)

Plate 6.7 Tewkesbury Abbey, Gloucestershire, surrounded by water,
23 July 2007
(Kirsty Wigglesworth/PA Wire)

Some 50mm or more of rain fell across a huge triangular-shaped zone stretching from Maidstone in the east to Bristol in the west to Shrewsbury in the north – an area twice as large as that affected by the late-June deluge. Totals exceeded 100mm over much of Oxfordshire, Gloucestershire, Worcestershire, Herefordshire, and small parts of Wiltshire and Warwickshire – an area of about 2,500 to 3,000 square kilometres – rather larger than the June deluges further north. However, because of the relatively low population density in the south-west Midlands, fewer people were flooded out here than had been in Yorkshire. In the last 60 years, the only high-summer downpours which matched this one both in volume and in geographical extent occurred on 25 August 1986 – the Bank Holiday washout associated with ex-hurricane 'Charley', 28 July 1969, 10 July 1968 and 12 August 1948. As a result of the storms of 10 July 1968, the 100mm threshold was exceeded at 65 rain-monitoring sites covering an area of 2,250 square kilometres extending from Devon to Lincolnshire, and 173mm fell, mostly within six hours, at Chew Stoke in Somerset. During the July 1969 event, some 5,250 square

Plate 6.8 Water, water everywhere. The River Thames at Pangbourne Bridge, here forming the boundary between Berkshire and Oxfordshire, at its greatest extent following the downstream flood peak on 24 July 2007
(Stephen Burt)

Plate 6.9 The Foudry Brook (a tributary of the Kennet), normally a small
stream, here almost a kilometre wide and submerging a field of maize,
near Stratfield Mortimer, Berkshire, following the heavy rainfall of
20 July 2007
(Stephen Burt)

kilometres of lowland Britain received over 100mm in less than 24 hours. All these downpours triggered destructive and widespread flooding. There were, of course, many other examples of similar or heavier rainstorms during other months of the year; these can be found in the chronicle in Part 2.

Once again, media reporting of the mid-July flood failed to place the disaster into a proper historical context. Once again we heard a location-specific statistic – that the water level in Gloucester was the highest since March 1947 – transformed into a general headline: 'Britain's worst flood for sixty years.' A shocking number of flooded householders were revealed to have no insurance; equally shocking were the vast quantities of household goods thrown away when all they needed was hosing down with clean water and drying out, while so many people who had endured water damage to their homes demanded, understandably but unrealistically, that the authorities provide immediate financial help. And in the most telling of ironies, some of the worst-hit areas of Worcestershire and Gloucestershire had their water supply cut off after a pumping station at Tewkesbury was inundated. Indeed, some 350,000 people were without water for several days, and the last were not reconnected until seventeen days after the flood first hit. Some 10,000 motorists were stranded on the M5, and 5,000 were stuck in Gloucester railway station as floodwaters rose on the night of the 19th–20th.

Taking the 2007 summer floods together, thirteen people died, 48,000 homes were damaged by floodwater, and 7,000 business premises were also flooded. Total insurance losses are estimated at between £2 and 3 billion.

It is tempting to describe 2007's holiday season as a 'retro summer' because in so many respects it was reminiscent of the many appalling summers that readers of a certain age will have been able to recall from the 1950s, 1960s and early 1970s. It is also interesting to note that summer floods occurred much more frequently during epochs (1912–31 and 1948–74) when summers were relatively cool.

Chapter 7

Destructive gales: two case studies

Measuring the wind

Wind is a rather contradictory element when it comes to measuring it. The direction from which the wind blows is easy to measure and even easier to estimate, and the very earliest meteorological instrument – the wind vane – provides us with a simple way of observing this particular parameter. Wind vanes and weathercocks have been popular for centuries, and for more than a millennium they have graced places of worship in many different parts of the world. The ancient Greeks thought of wind as the breath of the gods, and they named the wind directions after some of their deities: Zephyros, the west wind, Notos from the south, Apeliotes from the east, and Boreas was the north wind. In principle, the way we measure wind direction today is the same as it always was, though the vanes we use now are smaller and lighter and much more sensitive to slight variations in the movement of the air.

Wind speed, or wind force, presented early observers of the weather with a much tougher problem, one which was not solved until the end of the nineteenth century. The first step – and the easiest – was to estimate the force of the wind by reference to the behaviour of trees or the sea. There were probably many early efforts to systematize such a method, and the meteorological historian W. E. Knowles Middleton considered that Arab seafarers had certainly devised such a scale by the fifteenth century, perhaps much earlier. However, the one which survives to this day is the Beaufort Scale, drawn up in 1805 by a young naval officer, Francis Beaufort, who was the newly appointed commander of HMS *Woolwich*. His later career was very distinguished and he subsequently became

Admiral Sir Francis Beaufort, Hydrographer of the Navy, supporter of Admiral Robert FitzRoy, and mentor to Charles Darwin. Beaufort's original scale had fourteen categories (from 0 to 13) which referred to 'the speed a well-conditioned man-of-war would make and the amount of sail she could carry' (see Table 7.1).

Table 7.1 Beaufort's original scale

Force	Description
0	Calm
1	Faint air, just not calm
2	Light air
3	Light breeze
4	Gentle breeze
5	Moderate breeze
6	Fresh breeze
7	Gathering gale
8	Moderate gale
9	Brisk gale
10	Fresh gale
11	Hard gale
12	Hard gale with heavy gusts
13	Storm

Beaufort himself modified the scale slightly in 1806, but it did not catch on widely until FitzRoy adopted it in 1831, and rather tardily the Admiralty – perhaps grudgingly – gave it their imprimatur in 1838, by which time full descriptions of the sea state to be expected with each force were also annotated. An adapted version for use by meteorological observers on land and in coastal waters was devised in 1906 by a leading meteorologist of the time, Sir George Simpson (Table 7.2).

Such scales were important before the advent of reasonably accurate anemometers, and are still useful to observers who do not have access to an anemometer, but they are essentially subjective and provide only a crude assessment of the true wind force. Simple early instruments were designed to quantify the force exerted by the

Table 7.2 Beaufort's scale adapted by Simpson

Force	Description	Specification
0	Calm	Smoke rises vertically
1	Light air	Direction of wind shown by smoke
2	Light breeze	Wind felt on face. Vane moved by wind. Wind fills the sails of smacks
3	Gentle breeze	Leaves in constant motion. Wind extends light flag. Smacks begin to career
4	Moderate breeze	Raises dust and loose paper. Smacks carry all canvas, with good list
5	Fresh breeze	Wavelets form on inland waters. Smacks shorten sail. Small trees in leaf begin to sway
6	Strong breeze	Whistling heard in telegraph wires. Care required when fishing. Large branches in motion
7	Near gale	Whole trees in motion. Smacks remain in harbour. Those at sea lie to
8	Gale	Breaks twigs of trees. Generally impedes progress
9	Strong gale	Slight structural damage. Chimney pots and slates removed
10	Storm	Seldom experienced inland. Trees uprooted. Considerable structural damage
11	Violent storm	Very rarely experienced. Accompanied by widespread damage
12	Hurricane	

wind by fixing a broad metal flap along its upper edge, placing it in an exposed location, and allowing it to swing in the breeze – rather like those traditional pub signs found throughout Britain. Ideally, such a 'pressure-plate anemometer' should be attached to a wind vane to ensure that it is always facing the wind, and hinged loosely to minimize friction. The angle through which the plate is displaced by the breeze is proportional to the wind force, which in turn is roughly proportional to the square of the wind speed. The wind sock, still found alongside runways at airfields and airports, and more recently erected alongside those sections of roads and motorways which are particularly exposed to cross-winds, works on the same principle.

Descriptions and diagrams of the pressure-plate anemometer date back to the great flowering of scientific thought and invention in fifteenth- and sixteenth-century Italy, and England's own Renaissance Man, Robert Hooke, developed the idea further and produced a working example in 1667. In 1838 the pioneering Birmingham scientist and engineer, Abraham Follett Osler, constructed a self-registering anemometer (what we would now call an anemograph) which used the pressure-plate principle, and this is widely considered to be the first self-recording scientific instrument of any kind. These were installed at many observatories around Britain between 1840 and about 1900, and provided the first step forward along the road to accurate measurement of wind velocity. However, the instrument was often subject to mechanical failure, and as time passed it became increasingly evident that it could measure wind speed but crudely.

The wind-measuring instrument with which most of us are familiar – the cup anemometer – was invented in 1846 by Thomas Robinson, the third director of the Armagh Observatory in what is now Northern Ireland. Nowadays, such anemometers follow a three-cup design, but Robinson's had four cups. The difference in the pressure of the wind on the inside of the cup facing the wind compared with that on the outer surface of the opposite cup provides the force which causes the cups to spin, but the equation of forces is actually a complex one. Robinson believed that the ratio of cup speed to wind speed was a constant at 0.33, and experiments carried out at Greenwich and Kew Observatories seemed to bear this out. In practice, the relationship depended on the size of the cups and the length of the arms, and with the four-cup instrument it also varied with the actual speed of the wind. Later testing showed that the ratio for Robinson's design varied between 1.9 and 3.2, with an average of 2.2. The three-cup design, introduced in the 1920s, was found to have a much more nearly constant ratio; moreover, it responded faster to changes in the wind and was considered to provide a more accurate measurement of gusts.

Britain led the way in the development of meteorological instruments throughout the nineteenth century and for part of the twentieth, and it was in the UK that debate raged fiercest as to which design gave the truest representation of the actual wind speed, and which, if any, should be adopted as the official standard. Meteorological science had also been found wanting following the Tay Bridge Disaster in a severe gale in December 1879; nobody could say for sure how strong the wind had actually been. With this in mind, the Royal Meteorological Society established a committee in June 1885 to:

> investigate the relation between Beaufort's notation of wind force and the equivalent velocity in miles per hour, as well as the corresponding pressure in pounds per square foot, for each grade of the scale; to inquire whether any existing scale can be adopted or modified, and if not, to determine such equivalents as can be recommended for general and international use; also ... to report on the best mode available for the attainment of a satisfactory solution of the entire question of Wind Force.

That last all-encompassing instruction effectively asked the committee to choose a particular construction of anemometer as a standard for the foreseeable future. The Wind Force Committee met 28 times over a period of seventeen years, and it was the Society's great good fortune that a young meteorologist and instrument designer, William Henry Dines, was asked to join the committee in 1886.

Dines had, that very year, designed a completely different sort of anemometer. Usually called the 'pressure-tube anemometer', this instrument works on the principle that the pressure exerted by the wind on the air in a tube is proportional to the wind speed. In Dines's anemometer, the tube's opening is placed in the pointer of a wind vane, and is therefore always facing into the wind. The pressure in this tube is compared with the pressure in a second tube, open to the air but shielded from the wind, which acts as a control. The difference in pressure between the two tubes enables the wind speed to be represented, either on a dial, or, more usually, via a pen-arm on a chart – a self-contained, self-registering instrument, powered by the wind! The prototype was demonstrated to the Wind Force Committee on 18 December 1889 in London, experiments were carried out during the 1890s to verify its accuracy and to develop improvements, and, de facto, the Dines' PTA (as it was universally known) became and remained the standard instrument for measuring both wind speed and direction until the last quarter of the twentieth century.

This brief history of anemometry shows that there were very few reliable records of wind speed in the UK, or indeed anywhere in the world, until the very end of the nineteenth century. Although many gales of exceptional violence occurred during that century, and descriptive accounts can be found in the literature, it was not until the late 1890s that such accounts benefited from accurate measurements which in turn enabled comparisons to be made between different windstorms.

The gale of 26 January 1884

In his book *Historic Storms of the North Sea, British Isles, and Northwest Europe*, Professor Hubert Lamb identified twelve destructive gales during the twenty years from 1878 to 1897, of which six occurred between December 1879 and December 1886, but there were only five in the next twenty years. The early to mid-1880s, then, was an exceptionally windy period with much damage and loss of life both on land and at sea. It is no coincidence that the Royal Meteorological Society's Wind Force Committee was set up in 1885, for the perceived increase in the nation's vulnerability to high winds, especially in the wake of the Tay Bridge Disaster, had become a politically sensitive issue.

Arguably, the gale of 26 January 1884 was the worst of these: in one respect it was certainly *the* most extreme, since this was the occasion of the lowest barometric pressure ever recorded in the British Isles. At Ochtertyre in Perthshire a reading, adjusted to sea level, of 925.6 millibars was obtained at 9.45pm GMT, a figure which has not, before nor since, been equalled, and which has only once been approached.

The first half of the winter of 1883–84 had been exceptionally quiet, largely dominated by high-pressure systems, and there were several short-lived smog episodes in Britain's industrial conurbations. With anticyclones usually resident over southern Britain or the near-continent, there were very few days of easterly or northerly winds, so it was also a generally mild, frost-free winter with few significant snowfalls. Vigorous Atlantic depressions developed during the third week of January, tracking in a north-easterly direction between Scotland and Iceland, but the mid-January anticyclone proved remarkably resilient, maintaining a position just to the south of the British Isles, and this resulted in a marked steepening of the south-westerly gradient across north-west Europe from 19 January onwards. Between the 19th and 24th five intense, rapidly moving disturbances affected the northern half of the UK – those of the 19th, 20th, 21st and 24th passed to the north of Scotland, but that of the 23rd travelled across southern Scotland – bringing repeated but short-lived bouts of gale-force winds and periods of heavy rain.

This series of storms was described in detail in a paper read before the Royal Meteorological Society by William Marriott, the Society's Assistant Secretary. He noted that the first of these disturbances was a short-lived but intense feature, notable for the quite exceptional pressure falls (32.5 millibars in four hours at Stornoway on the Isle of Lewis), and a sustained wind recorded over a 65-minute period at Sandwick, Orkney, of 92mph. Although this figure is regularly quoted, it should be noted that it was recorded on a Robinson anemometer with the incorrect adjustment factor. Applying a correction for this, and another for the period over which the record was made (the standard period for measuring mean wind speed is 10 minutes, while gusts are measured over 3 seconds), gives an estimated maximum mean wind speed of 72mph with gusts probably between 100 and 110mph.

Atlantic charts were not prepared routinely until well into the twentieth century, largely because the transmission and collection of ships' observations had not yet been systematized, although it was often possible – painstakingly, and with considerable consumption of time – to construct such charts after the event. However, the gradual digitization of the vast volumes of historical data in recent decades has enabled, in a project which was launched in 1991, pressure charts to be constructed for much of the northern hemisphere back to 1880, and for Europe, the north Atlantic and North America this ongoing exercise should eventually take us back even

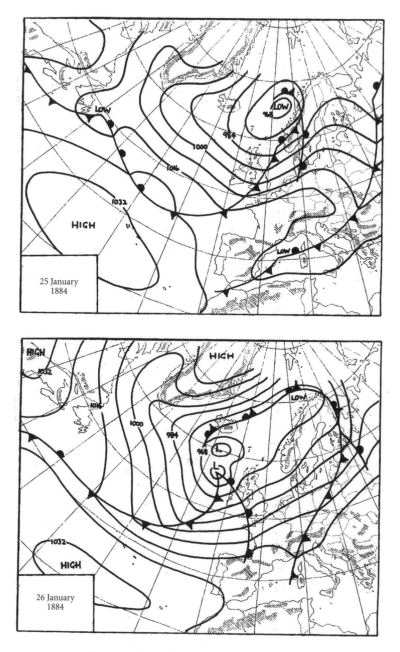

Figure 7.1 Synoptic charts, 25–28 January 1884

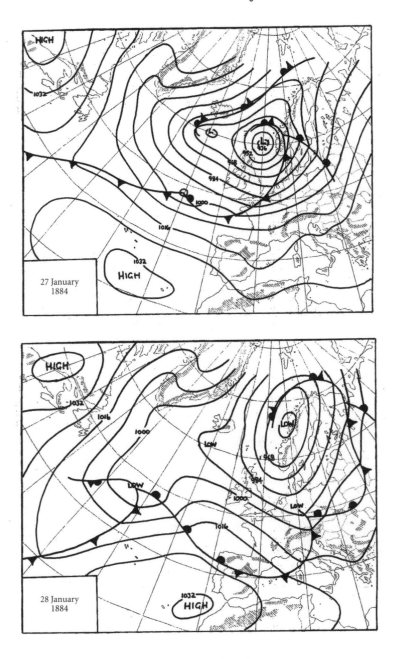

Figure 7.1 continued

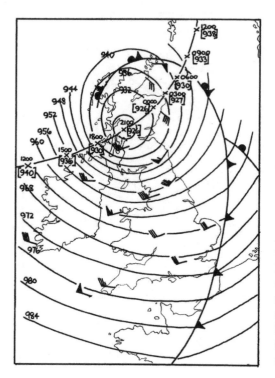

Figure 7.2 Detailed weather chart for the British Isles for 9pm on 26 January 1884, also showing the depression track between noon on the 26th and noon on the 27th

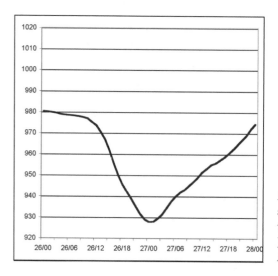

Figure 7.3 Constructed barograph trace for 26–27 January 1884 at Aberdeen

further. Even so, these reanalysed pressure charts are based on 5-degree latitude/longitude grid, so they can only give a broad-brush picture. But with the additional help of an experienced meteorologist's eye, we can now view the entire life history of the record-breaking depression which crossed Britain on 26 January 1884. It probably began life as a shallow disturbance on a trailing cold front over the eastern USA on the 23rd, although it may have contained the remnants of an earlier depression which lay near Chicago on the 21st/22nd. By the morning of the 25th it was centred just east of Newfoundland, still quite innocuous, with a central pressure of 1011 millibars, but from here it raced across the Atlantic, deepening rapidly, to be about 1,000km west of Galway on the west coast of Ireland by the early hours of the 26th, with a pressure of about, or perhaps rather below, 950 millibars at its centre. The dramatic deepening and rapid movement are classic symptoms of a powerful jet-stream in the upper troposphere, and we may surmise that there existed aloft a west-south-westerly jet blowing at 220mph or more. By midday on the 26th the centre of the surface depression lay off the coast of County Mayo at about 940 millibars, whence it tracked to Islay at 930 millibars by 6pm, and to Aberdeenshire at 926 millibars by midnight. By midday on the 27th the centre lay just east of Shetland at 938 millibars, and by the same time on the 28th over the central Norwegian coast at 955 milibars, where it gradually filled up in situ during the 29th and 30th.

The strongest winds were recorded in three phases: in the southerly airflow on the forward side of the depression, ahead of the feature we should now identify as a warm front; along a very active and well-marked squall-line which also brought thunder, lightning and hail, something we should now call a cold front; and in the zone of steepest pressure gradient for westerly winds, some 200–400km south of the depression centre, a configuration we might now call a 'sting-jet'.

Mean hourly wind speeds were recorded at fifteen sites equipped with Robinson anemometers around the British Isles. Tables 7.3 and 7.4 show the raw observations followed by corrected figures for highest mean hourly, maximum ten-minute, and maximum gust, at a selection of these weather stations.

Table 7.3 Wind gusts at selected weather stations – raw observations

	Reported max hourly (mph)	Corrected max hourly (mph)	Corrected max 10-min (mph)	Estimated max gust (mph)
Greenwich (London)	58	43	47	77
Yarmouth (Norfolk)	47	34	37	54
Falmouth (Cornwall)	63	46	51	74
Isles of Scilly	56	41	43	57
Holyhead (Anglesey)	70	51	54	71
Alnwick (Northumberland)	76	56	62	90
Aberdeen	59	43	47	69
Sandwick (Orkney)	55	40	42	56
Armagh	69	51	56	82
Valentia (Co. Kerry)	68	50	53	70

These figures demonstrate the great geographic extent of the gale, and the exceptional ferocity of it over much of the northern half of Ireland, northern England, and, by inference, southern Scotland, but they also show that the fiercest winds were not particularly long lived. As a rule of thumb, meteorologists consider sustained winds of 40mph or more, and gusts in excess of 70mph, to be powerful enough to cause structural damage to buildings, to uproot healthy trees, and to cause severe problems to transport. From the estimation that gusts exceeded 80mph at Armagh and may have reached 90mph at Alnwick, we might infer that there was widespread damage and dislocation in these regions, with the likelihood

Table 7.4 Hourly mean wind speeds at selected weather stations –
corrected to present-day standards

26th January Hour ending (GMT)																27th January								
09	10	11	12	13	14	15	16	17	18	19	20	21	22	23	24	01	02	03	04	05	06	07	08	09

Corrected hourly wind speed (mph)

Greenwich

14	13	14	17	13	22	30	37	43	43	43	32	24	26	29	21	23	20	24	21	18	19	22	19	21

Falmouth

23	31	33	37	45	43	46	46	32	29	26	25	28	26	21	18	17	19	17	20	23	22	21	23	26

Holyhead

29	35	43	43	41	51	48	40	44	40	38	39	40	39	38	39	39	38	32	37	37	34	34	31	25

Alnwick

11	11	14	14	17	26	29	31	31	26	18	22	41	52	56	56	51	47	44	45	43	47	42	32	34

Aberdeen

6	7	7	8	10	22	31	39	43	34	31	28	30	28	13	11	14	18	17	20	28	31	33	37	40

Armagh

21	20	27	30	37	36	37	46	51	39	35	28	25	27	28	25	27	21	23	18	14	13	14	14	13

Valentia

47	46	50	48	41	41	34	33	26	26	30	30	30	25	27	22	20	17	24	26	29	29	30	26	25

of some loss of life. One further point to emerge from the tabulation is how sharply the winds diminished either side of midnight at Aberdeen as the depression centre passed just to the north of the city, with several hours of gale-force south- to south-easterly winds ahead of it, and a renewed gale, but this time from the north-west, in its wake.

The observatory on the summit of Ben Nevis had been completed the previous year, and weather observers had been resident there since June 1883. Although the depression track lay to the south of the Ben, the weather there was sufficiently bad for the duty observer, Mr Omond, to pen an account of the storm which was published in *The Scotsman* a few days later:

> The lowest [barometer] reading, 23.173 ins [784.7 millibars] occurred at 8.30 pm, while about the same time the lowest reading at Fort William (the low level station), reduced to sea level, was 27.467 ins [930.1 millibars] or 4.294 ins higher than at the top of the Ben.
>
> At noon the temperature on the summit of mountain was 15°[F, −9.4°C], but from that time until 10 pm no observations could be taken on account of the fury of the gale. At 10 pm the temperature was 22°[F, −5.6°C]. The direction of the wind at noon was south-east, force 10; at 7 pm it was still blowing from the south-east, force 8; but at 10 pm, after a brief calm, it shifted to east-north-east, force 4. During all this time snow and fog with heavy drift were prevalent. While the barometer was falling most rapidly, Mr Omond took observations every quarter of an hour, though with considerably difficulty on account of the 'pumping' of the mercury, the violence of the gale as it passed over the observatory causing a constant pulsation of the air inside the room. The pumping amounted to about 0.1 in [3.4 millibars]. Had the mercury gone down 0.2 in lower it would have been beyond the scale of the instrument, although this was specially constructed for the station. The reading of the barometer on October 16th, 1883, the day the observatory was opened, was 28.944 ins [980.2 millibars]. This gives an observed range of 5.771 ins [195.5 millibars] at that level.
>
> In connection with the thermometer readings, it may be mentioned that at 1 pm Mr Omond made an attempt to get at the [thermometer] screen. Tying a rope round his waist, the end of which was held by an assistant within the porch, Mr Omond crept cautiously out from the shelter of the observatory, but so great was the violence of the gale that he could make no headway against it, and was glad to return. At 7 pm another attempt was made. The observers got as far as the screen, but found it impossible to read the thermometers owing to the blinding drift lashing in their faces. At 10 pm it was calm enough for Mr Omond to go out alone, and the reading of 22°, recorded above, was obtained.

The severe gale was remarked upon by official weather observers in all corners of the country. The following are but a selection:

London: Violent SW gale with heavy rain.

Hythe (Kent): From 5.30 to 9pm terrific hurricane, with heavy thunder, rain and hail squalls.

Rodmersham (Kent): High wind sprang up suddenly in the early afternoon, increasing to a hurricane towards night, blowing in terrific gusts between 6 and 8pm . . . heavy rain fell, and there was frequent lightning; much damage was done.

Littlehampton (Sussex): Awful gale; a large pane of glass blown in from a window that had resisted all previous gales.

Babbacombe (Devon): Fearful SW gale; rain, thunder and lightning.

Melton Mowbray (Leicestershire): Terrible gale 4 to 8pm.

Mansfield (Derbyshire): Very violent gale, with snow, vivid lightning and thunder.

Hull (East Riding): A heavy gale commenced somewhat suddenly about 5.30pm, blowing with intense force from the SW until 7pm, when it rapidly subsided almost to a calm, until about 9pm when it again rose, and continued with hurricane force throughout the night.

Morpeth (Northumberland): A great gale with barometer very low.

Carlisle (Cumberland): A terrific hurricane, dislodging slates from several houses in the town, blowing down several chimneys and five headstones in the cemetery.

Llandovery (Carmarthen): The heaviest gale remembered, from 1pm to 5pm.

Garliestown (Wigtownshire): Very severe storm of wind, which swept down thousands of trees in the district, the ground at the time being saturated with water.

Melrose (Roxburghshire): Hurricane from west, strongest between 2 and 4am on 27th; many trees uprooted, houses injured, and several tombstones blown down in the Abbey burying ground.

Cassillis (Ayrshire): Storm of extraordinary severity, with snow and rain.

Aviemore (Inverness-shire): Severe gale from the west; highland railway blocked by snow at night, and the telegraph line much damaged.

Tuam (Co. Galway): Violent storm, pressure remarkably low, many trees being blown down.

Enniscoe (Co. Mayo): A perfect hurricane.

Enniskillen (Co. Fermanagh): A very severe storm, with remarkable fall of pressure; old trees which had stood for 130 years were blown down.

Newtownards (Co. Down): The most terrific storm since 1839; wind 78 miles an hour.

Bellarena (Co. Londonderry): A great hurricane from SW, W, and NW, which blew down a great number of old trees.

Buncrana (Co. Donegal): Gale of unusual force, commencing at 6pm and continuing with unabated force till 3am on 27th, causing much damage on land and sea.

Curiously, the accounts of major weather events in the scientific literature during the late nineteenth century largely ignored the human effects of the disasters, preferring to concentrate on describing in detail the meteorological circumstances which caused them, which in turn meant collecting, examining and analysing all observational data. However, it has been possible to piece together some of the results of the gale: there was extensive damage to forests and other woodland in all parts of the UK, more than one million trees were uprooted or otherwise broken on the Cargen estate, south of Dumfries, and a 600-year-old yew tree in Hemsworth churchyard in the West Riding of Yorkshire lost several of its principal branches; thousands of buildings, especially in Wales, northern England, northern Ireland and southern Scotland, suffered damage to roofs, chimneys and windows; and hundreds of miles of telegraph wires were brought down. The grand pavilion at Llandudno, just completed, lost its domed roof, but repairs were expedited during 1884, and the building was to last for over a century before becoming a victim of arson in 1994. Blackrock Castle Observatory, on the banks of the River Lee, 2km to the east of Cork city centre, also lost its dome, while debris from a partially demolished windmill at Lakenham, near Norwich, was deposited up to 250 metres away by the wind. There were many maritime losses around Britain's coastline too: the *Simla* sank after a collision with the *City of London* off the Isle of Wight, with the entire crew of 20 drowned; the trawler *Foxhound* out of Hull foundered in the Humber estuary with the loss of all hands; the *Victoire* out of Belfast sank in Loch Ryan, just north of Stranraer; the fishing smack *Quiz* out of King's Lynn also sank and the crew of the Spanish brigantine, *Juan de la Vega*, bound for Porthcawl with a cargo of pit-props, were rescued by lifeboat.

We have already noted that the effects of the gale were felt over practically the whole of the British Isles; in fact, they were felt much more widely, with Beaufort force 9 or 10 winds observed on the other side of the North Sea in southern Norway, Denmark, north-west Germany and the Netherlands, in the southern Baltic, and well inland in central and southern Germany, Switzerland and over much of France. At Lorient on the coast of Brittany the wind reached force 12 for a time. Before he built the eponymous tower, Gustave Eiffel's most prestigious project was the design and construction of the famous Garabit railway viaduct over the Truyère river near Ruynes-en-Margeride, in the Massif Central. This dramatic structure was approaching completion in January 1884, but the high winds on the 26th and 27th demolished parts of the viaduct, and delayed the opening of the single-track railway line which it carried until later in the year.

It is practically impossible to calculate the loss of life due to severe weather events before the twentieth century, but we may surmise, from the available accounts of losses at sea and scattered reports of deaths on land, that the death toll in the British Isles and home waters in all likelihood exceeded 50, but was probably fewer than 100.

The gale of 19 January 2007

There was a sharp increase in the frequency of damaging winter gales in the UK and other parts of north-western Europe between the years 1987 and 2002, which mirrored the general rise in the frequency and intensity of westerly winds during the winter half-year over the same period. There has been something of a decline in the frequency of both elements after 2002, and the winter of 2005–6 had been remarkably free of strong winds. It was, in fact, the quietest winter since at least 1991–92, and arguably since 1963–64. Vigorous Atlantic depressions returned during the late autumn of 2006, and the period from mid-November until mid-January (with a ten-day break either side of Christmas) saw a succession of active disturbances sweeping across the British Isles from the west bringing

Plate 7.1 The Golden Mile as your rarely see it: severe gales and rough seas
batter the sea front at Blackpool, 18 January 2007
(John Giles/PA Wire)

frequent bouts of very wet and very windy weather. The depression
responsible for the severe gale of 19 January was the last in the succession, and in its wake high pressure settled over the country for
almost three weeks, providing quiet conditions for clearing up after
the damaging winds.

The depression itself had its origins some six days earlier in a
marked trough which was slow moving over the American mid-
west. A cold front lay in the trough, defining the boundary between
very warm and very moist air moving northwards from the Gulf of
Mexico, and much colder and drier air moving south from the
Canadian Arctic. Heavy rain and snow associated with the trough
fell for several days over a wide area in the eastern half of the USA.
A localized fall in pressure occurred on the 14th on the western
flank of the Appalachians, resulting in the formation of a small
depression, and this feature deepened significantly over the next 24
hours as it travelled slowly north-east, so that a cyclonic circulation was centred over Ohio at 1009 millibars at midday GMT on

the 15th. It now lay below the right-hand side of the entrance of a powerful west-south-westerly jet-stream which had been controlling the passage of depressions across the Atlantic Ocean for many days. Forecasters call this a 'development area' because the dynamics of the jet-stream cause a surface depression under the right entrance to intensify sharply, and this is exactly what our 'low' did, accelerating across the ocean at the same time. The location of the depression centre at twelve-hour intervals was as shown in Table 7.5.

There are several striking characteristics of this particular disturbance: the explosive deepening late on the 16th and early on the

Table 7.5 Location of depression centre at 12-hour intervals

Date	Time (GMT)	Central pressure (millibars)	Location	Speed of movement	Rate of deepening
15th	1200	1009	Ohio		
16th	0000	1004	Cape Cod	46 knots	5mbar/12hr
16th	1200	999	Nova Scotia	48 knots	5mbar/12hr
17th	0000	984	48N 47W	55 knots	15mbar/12hr
17th	1200	968	51N 38W	43 knots	16mbar/12hr
18th	0000	967	54N 20W	50 knots	1mbar/12hr
18th	1200	966	Central North Sea	50 knots	1mbar/12hr
19th	0000	963	Kaliningrad	45 knots	3mbar/12hr
19th	1200	963	Just west of Moscow	30 knots	0mbar/12hr
20th	0000	965	Just north of Moscow	20 knots	−2mbar/12hr
20th	1200	976	60N 42E	25 knots	−11mbar/12hr
21st	0000	982	64N 52E	30 knots	−6mbar/12hr
21st	1200	982	68N 61E	30 knots	−1mbar/12hr

17th, the maintenance of its intensity for almost 72 hours thereafter, its speed of movement across the Atlantic, traversing the ocean in less than two days, and its long life – it continued its east–north-eastward journey across northern Russia and into the Arctic Ocean, and it still retained a circulation north of the Siberian coastline into the 25th. It is important to note one other aspect of this depression: in contrast to the record-breaking 'low' of 26 January 1884, this one achieved no exceptional depth. Cyclonic centres of 960–970 millibars over the Atlantic are two a penny at this season, and several of them can be found in the vicinity of the British Isles in most winters. The ferocity of the gale associated with this particular feature was the consequence of the dramatic tightening (or steepening) of the pressure gradient between the low centre and an extensive high-pressure area lodged over the Iberian peninsula which, curiously, also reached its greatest intensity – 1038 millibars – during the 18th–19th.

This facet of the gale is emphasized by the fact that, over the British Isles, no strong winds were reported to the north of the depression centre. By contrast, on the southern flank of the 'low' there was a progressive increase in wind speed during the morning, followed by a sudden increase in the gust ratio at the peak of the gale during the middle of the day, resulting in some powerful gusts at this time which were responsible for much of the reported damage, after which there was a progressive decrease in both mean wind and gusts later in the afternoon and evening. The marked increase in gustiness was associated with the passage of the system's cold front; the introduction of slightly colder but much drier air triggered dynamic instability which resulted in increased turbulence. Although this was treated as some sort of dramatic revelation in some media reports at the time, it is actually a common feature of many westerly gales in the UK. A further minor peak in wind speed was noted in some southern counties of England overnight on the 18th–19th; this was associated with a small secondary disturbance which travelled rapidly eastwards across southern England, thence towards the Low Countries.

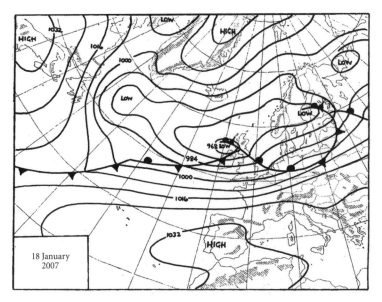

Figure 7.4 Synpotic chart for 18 January 2007

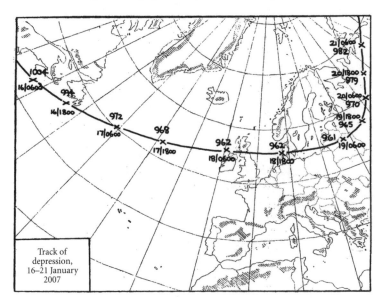

Figure 7.5 Depression track between 16 and 21 January 2007

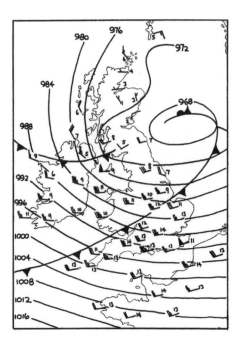

Figure 7.6 Detailed weather chart for noon on 18 January 2007

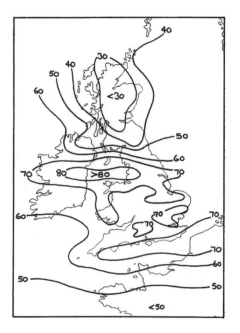

Figure 7.7 Maximum gust speed (in mph over the British Isles and northern France on 18 January 2007

Table 7.6 Maximum recorded gusts in the UK and Irish Republic,
18 January 2007

Location	mph
Capel Curig (Caernarfon)	85
Dublin airport	85
Crosby (Merseyside)	84
Squires Gate (Lancashire)	84
Heathrow airport	78
Marham (Norfolk)	78
Odiham (Hampshire)	78
Sheerness (Kent)	78
Wittering (Cambridgeshire)	78
Church Fenton (Yorkshire)	77
Lyneham (Wiltshire)	77
Northolt (London)	77
Langdon Bay (Kent)	76
Lee-on-Solent (Hampshire)	76
Manchester airport	76
Mumbles (Glamorgan)	76
Ronaldsway (Isle of Man)	76
Aberporth (Ceredigion)	75
Liverpool airport	75
Southend (Essex)	75
Valley (Anglesey)	75
Waddington (Lincolnshire)	75
Boscombe Down (Wiltshire)	74
Gravesend (Kent)	74
Hawarden (Flintshire)	74
Humberside airport	74
Leeds/Bradford airport	74
Scampton (Lincolnshire)	73
Wattisham (Suffolk)	73
Knock airport	71
Pembrey (Carmathen)	71
Thurleigh (Bedfordshire)	71
Benson (Oxford)	70
Casement (Dublin)	70
Cranwell (Lincolnshire)	70

(The widely reported figure of 99mph at The Needles lighthouse, Isle of Wight, was obtained at a non-standard site, and probably exceeded the true wind by approximately 20 per cent.)

The gale was just as severe on the other side of the North Sea. Gust speeds at low-level sites peaked at 81mph in Belgium, 83mph in the Netherlands, 89mph at Düsseldorf airport in Germany, 85mph in Poland and 96mph in the Czech Republic. High-level sites were much windier, and 126mph was recorded at Wendelstein in the Black Forest and 123mph on the summit of the Brocken, both in Germany.

The death toll in the UK was nineteen, but a total of at least 60 people lost their lives across all of northern and central Europe, with several fatalities reported from France, the Netherlands, Germany and the Czech Republic, as well as from the UK. Transport in Britain and home waters was thrown into chaos. A British container ship, the 62,000-tonne *Napoli*, got into difficulties 120 kilometres south of The Lizard, and the 26 crew had to be rescued by helicopters from

Plate 7.2 Police guide traffic around several overturned lorries on the A1, just outside Newry, County Down, in Northern Ireland, 18 January 2007
(AP Photo/Peter Morrison)

Table 7.7 Ten-minute mean wind speed, plus highest gust, at each hourly observation at selected sites (in mph), 18 January 2007

Time (GMT)	Dublin airport	Crosby	Marham	Lyneham
00	12	29/35	18	12
01	18	18/35	18	18
02	12	18	18	12
03	18	18	18	12
04	12	18	18	18/29
05	29/44	18	18	29/39
06	29/38	29/43	18	29/38
07	35/60	29/37	29/37	29/42
08	35/45	46/66	29/38	29/38
09	35/45	—	35/50	29/45
10	41/62	41/58	41/58	35/48
11	52/85	41/62	35/50	35/62
12	46/68	46/70	35/50	35/55
13	46/74	—	35/52	41/77
14	41/54	46/75	46/78	46/68
15	35/45	52/84	41/57	41/61
16	29/38	46/73	46/73	41/68
17	23/34	41/63	41/63	35/53
18	29/37	35/54	41/68	29/44
19	29/38	35/45	35/65	29/47
20	29/36	35/45	29/50	29/46
21	23	29/40	18	29/37
22	23	35/40	23	29/40
23	29/36	29/44	23	18
24	29/36	29/38	18/39	18

Plate 7.3 A sight repeated hundreds, maybe thousands, of times across the country; an ash tree brought down by winds gusting to 70mph at Whipsnade, Bedfordshire, just misses the author's home, 18 January 2007
(Philip Eden)

the Royal Naval Air Station at Culdrose in south Cornwall. Glass roof panels fell onto the concourse at London Bridge station, triggering its closure for several hours, and the roof of one of the stands at Lord's cricket ground was demolished, trailing debris across the ground; fortunately no one was hurt at either location.

Over 300 outbound flights were cancelled at British airports, including 192 at Heathrow alone. On the railways, the Eurostar service to Paris and Brussels was suspended for a time, the main East and West Coast lines from London to Scotland were on emergency timetables and operated at reduced speed for much of the day, as was the Great Western line to Bristol and Cardiff. Rail services did not fully return to normal until the following Monday (the 22nd). Hundreds of roads, including some sections of motorway, were closed for a variety of reasons; there were hundreds of uprooted trees, fallen power and telephone lines, high-sided vehicles which had been blown over, as well as a number of traffic accidents. There was also some flooding, notably in north Wales and in southern

Scotland, and heavy snowfalls disrupted transport in central and northern Scotland. Ferry services to France, Ireland, Belgium, the Netherlands and Denmark were cancelled for most of Thursday and part of Friday too. Approaching 150,000 homes were without electricity for a time, and although power supplies were generally restored before the end of the day, some homes in Wales remained without power until Sunday the 21st.

Unlike the 1884 gale, the 2007 event was well predicted by the various forecasting agencies and companies. The forecasters, both in the Meteorological Office and in the more prominent private-sector organizations, did an excellent job with timely and accurate warnings during the three or four days leading up to the event, having originally flagged up the possibility of high winds about a week in advance. It is a moot point, though, whether such warnings, good as they were, actually diminished the number of lives lost, or even the economic costs of the storm.

An ugly feature of severe weather reporting in the press and the broadcast media which has developed rapidly in the last decade or two, was well to fore on this occasion. Newspapers, especially those at the popular end of the spectrum, love sensationalizing weather stories, but for most of the twentieth century the sober scientists who occupied the forecasting offices refused to play ball, preferring simply to supply unadorned facts and figures. If the papers wanted sensation, they had to make it up themselves. That all seemed to change from about 1990 onwards with the advent of public relations departments, press offices and the importance of 'profile', whatever that may mean.

Thus on 18 January, in a crass attempt to feed the news industry's insatiable appetite for exaggerated headlines, the Met Office's press office went through contortions to find an eye-catching statistic. Because the recorded winds, both sustained speeds and peak gusts, were not truly exceptional – indeed higher gusts had been recorded in many places just a week before – they finally came out with this: 'The last time such high wind speeds were experienced across such a large area of the UK was . . . on 25 January 1990'. That is where the 'worst for 17 years' tag came from, heard so widely in radio and TV new bulletins in subsequent days.

Even that is arguable. Similar wind speeds were reported on 20 March 2004 over a rather wider area of England and Wales, with a peak gust of 101mph in north Wales. Stronger winds hit a larger area during a pair of windstorms on 7–8 January and 11–12 January 2005, although on those occasions it was the northern half of the UK which suffered most. Most national newspapers and broadcasters are London-based, and many (though not all) of them are notoriously metrocentric in their outlook, so northern Britain tends to matter less to them. Ignored, too, were those many ferocious gales between 1990 and 2006 which cost dozens of lives (over 100 during the whole of that period), damaged thousands of properties, uprooted hundreds of thousands of trees, and left an aggregate of almost 2 million people without electricity for days – ignored because they did not affect 'such a large area of the UK'.

Without in any way being an exhaustive list, we might mention those of 13, 17, 23 and 24 January 1993; 8–9 December 1993 which hit southern Britain badly; Christmas Eve 1997, following which tens of thousands of people in northern England and north Wales endured a joyless Christmas with no electricity for four days; Boxing Day 1998 with gusts to 103mph and six weather-related deaths; Christmas Eve 1999; 28 January 2002; 22 February 2002, when the roof of York Station collapsed; and finally 26 October 2002, which left thousands without power for almost a week in East Anglia and the Midlands.

There was a change at the head of the Met Office's press office within weeks of this low point in its history, and it is satisfying to note that, under new leadership, its output appeared to revert to sound explanations supported by accurate facts and figures.

Chapter 8

Drought: two case studies

Drought and the birth of the British Rainfall Organization

An affable young man from Pimlico strolled into Cavendish Square on 25 March 1856, rang the bell of what appeared to be a private house, and was ushered into a small sitting-room where about a dozen gentlemen were seated around a large table. The young man was George James Symons, barely seventeen years old, and the meeting was of the British (now the Royal) Meteorological Society which had been formed six years before. This was the meeting at which the precocious Symons was elected a Fellow of the Society, a Society which he was to serve as a council member for 37 years, as Secretary for 24 years, and as President twice, in 1880–81 and in its Golden Jubilee year in 1900. Sadly he suffered a stroke in February 1900, died four weeks later, and was present only in spirit at the suitably muted jubilee celebration.

Symons had shown a keen interest in the natural world from a very early age, and as a boy kept a detailed weather diary. His scientific studies coalesced around meteorology, he read several papers on thunderstorms and associated phenomena to the Society while still a teenager, and he finally gained employment under Admiral Robert FitzRoy at the newly formed Meteorological Department of the Board of Trade (the forerunner of the Meteorological Office) in 1860.

Symons's burgeoning meteorological curiosity coincided with a succession of very dry years, to the ultimate good fortune of British meteorology. The year 1852 had been the wettest for almost a century, but the next seven years were all drier than average, and three of them – 1854, 1855 and 1858 – were among the driest on record. The mid-nineteenth century was a period of rapid urbanization in Britain, and one of the most difficult problems in the quickly growing towns and cities concerned the supply of water. The dry years of the 1850s led to a widespread fear that the climate was actually getting drier and that existing water supplies might fail, with catastrophic results for public health, especially among the poorer classes. The Superintendent of the Meteorological Department at the Greenwich Royal Observatory fuelled the flames by demonstrating a progressive decline in the rainfall in London over the previous 45 years; he should have known better than to draw general conclusions for the whole country from a particular series of rainfall observations at a single site in south-east England, but there were no published records representative of the entire nation. In 1859, Field Marshal George Hay, the Eighth Marquess of Tweeddale, in his capacity as President of the Scottish Meteorological Society, proposed a cash prize for the best essay on the subject of the recent rainfall deficiency, with special reference to Scotland. The prizewinner, Thomas Jamieson, painstakingly collated data from over twenty stations and he concluded that there was no underlying downward trend. Jamieson was no ordinary essayist; at the time he won the prize he was the estate manager at Ellon Castle in Aberdeenshire, but he was already a keen geologist and a few years later became a lecturer in geology at Aberdeen University. The 1850s had certainly delivered quite a scare, and water supply was never again taken so much for granted as it had been before the dry period.

Symons's particular inquiry into the meteorology of thunderstorms naturally included an analysis of rainfall patterns associated with different kinds of storms, and his activities therefore soon spilled over into a more general interest in the measurement of rain. When he came to look at what information was available, he was immediately struck by the lack of uniformity in observational practice, in instrumentation, and in the siting of gauges, and he was also disappointed to find that no one had yet attempted to collate rainfall observations, either in real time or in archived form. He took his concerns to FitzRoy, but the admiral was not much interested in rainfall – it is not that important at sea, after all – and he firmly discouraged Symons from pursuing his new interest during office hours. Symons was not so easily diverted, however, so he used his spare time instead, and roped in his mother as his chief assistant.

In his quest for data for his analysis of thunderstorms, Symons had already made written contact with dozens of members of the

British and Scottish Meteorological Societies, and in 1860 and 1861 he wrote to all the rainfall observers he was aware of around the country, informing them that he had begun the 'Herculean labour of collecting the published and unpublished results' of all those observers who were prepared to contribute. His first volume – a slim pamphlet, really – was entitled *English Rainfall* and contained the records from 168 observers, but as more and more data rolled in from his respondents he republished the 1860 records along with those for 1861 under the title *British Rainfall*, beginning a series of annual volumes which continued until 1991.

The extended 1860 tables contained 424 records; there were over 1,000 by 1865, and 1,500 by 1870, by which time *British Rainfall* had grown into a sizeable tome 184 pages long. His work did not simply involve collating contemporary records. He immediately understood that for a national network to be of any value to researchers and future historians, the method of observation had to be standardized and regulated. A sizeable minority of the existing rainfall observers, naturally convinced that their own methods were the right ones, did not take kindly to being told what to do. Somehow, through a combination of diplomacy, cajoling, persuasion and a simple appeal to good scientific practice, Symons succeeded in carrying the vast majority with him.

In addition to developing and running this real-time network of observing sites, Symons was also collecting old observations extending back as far as 1677, and his growing obsession quickly became a full-time job, so he severed his connection with the Board of Trade and Admiral FitzRoy in 1863. Money was now a problem. The Symonses had never been more than modestly well off, and after his father's death in about 1855 the family's circumstances became increasingly straitened. He had earned a good salary under FitzRoy, but he now had to eke out an existence by requesting donations from his observers: by requesting a fee for testing and calibrating gauges, by charging five shillings – a princely sum in those days – for the annual volume, and by soliciting grants from scientific bodies such as the Royal Society, the British Association and the Meteorological Council. He also began publishing a monthly *Rainfall Circular* in 1863, and this developed into *Symons's Meteorological Magazine* from 1866 onwards, for which he charged a small subscription. The magazine contained all the meteorological news of the day, accounts of interesting weather events, and vibrant letters pages. Like *British Rainfall* it ran until the early 1990s. Rarely can one man have been responsible for the launching of two important periodicals which lasted for a century and a quarter, having run them single-handed for the first 35–40 years.

Under Symons's leadership, then, and with precious little help from 'official' science, the United Kingdom developed the first serious rainfall network in the world, and one which was to lead the world until financial expediency, administrative indolence and changing technology provoked a progressive atrophy from the 1980s onwards. By the time of Symons's death in 1900 the network of observers was 3,500 strong, and it continued to grow during the first half of the twentieth century; there were 5,000 of them in 1919 when the British Rainfall Organization was taken over by ('merged with' was the wording used at the time) the Meteorological Office, and as many as 6,600 for a time between 1965 and 1974. When the last annual publication was made in 1991, however, the number had declined to 4,000, and by the early years of the new century to little more than 3,000.

But it was the coincidence of a worrisome drought and a young man's burgeoning obsession that gave us such an amazing scientific resource. And if the drought did not last, the obsession did, and it enabled those of us who came after to examine in great detail the progress of future droughts, and to be able to place them in proper historical and statistical contexts.

The drought of 1887

Averaged over England and Wales, 1887 was the driest calendar year of the entire nineteenth century. The most acute phase of the drought was during the spring and early summer and was exacerbated by recurrent heatwaves and abundant sunshine during late June and July. Rainfall continued below average during the first half of 1888, although there was generally just sufficient rain during these months to prevent the drought reaching calamitous proportions. The water shortages were ended abruptly in July 1888 which was one of the coldest and wettest summer months ever recorded in the UK, and there were further heavy downpours during the subsequent August and November.

One way of assessing the rarity of dry weather over different

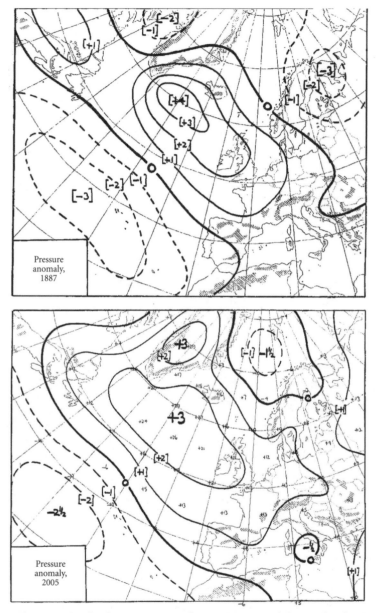

Figure 8.1 Sea-level pressure anomaly over Europe and the north Atlantic during the two drought years, 1887 and 2005 (expressed as millibars' difference from the mean pressure for the period 1961–90)

Table 8.1 Monthly rainfall over England and Wales,
October 1886 to December 1888

Month	Rainfall (mm)	Percentage of 1971–2000 normal
1886		
October	135.6	146
November	99.4	99
December	145.2	144
1887		
January	71.8	76
February	24.1	36
March	49.8	68
April	36.6	59
May	54.0	88
June	20.8	30
July	38.1	66
August	53.4	74
September	94.2	121
October	71.4	77
November	86.1	86
December	69.0	68
1888		
January	39.0	41
February	38.9	58
March	97.2	133
April	49.8	80
May	35.7	58
June	73.9	108
July	156.6	275
August	86.3	119
September	34.2	44
October	43.5	47
November	150.1	150
December	73.1	72

periods is to calculate the 'return period', which may be defined as the average time between occurrences of a similar event. Thus, if we have 100 years of records, and lower rainfall for a particular period has occurred only five times, the return period is 100/5, or 20 years. Normal practice is to fit the rainfall data to a standard statistical distribution, because this irons out any small inconsistencies that may exist in the actual records. When talking about return periods, it is important to be very clear and specific about what is being compared: a very dry calendar year like 1887 has a return period of over 100 years when compared with other calendar years, but the likelihood is that the driest weather straddles two calendar years. We therefore need also to calculate the return period for all twelve-month spells, no matter which was the starting month. Thus we need to distinguish between return periods for the particular months under consideration, and for similar groups of months at other times of the year.

The shortage of rain first came to prominence towards the end of February. In the late nineteenth century February was usually referred to, in any discussion of weather or climate, as 'February Fill Dyke', and the assumption was that February was a notoriously wet month and could be relied upon to fill the 'dykes' – or ditches. During the twentieth century the received wisdom was that the ancient expression was actually an exhortation – a request to the Almighty to deliver some rain in February – so that there was plenty of water in the ditches at the beginning of spring, a season known, at least in lowland Britain, for its low rainfall and drying easterly winds. Anyway, February 1887 was an exceptionally dry month with less than 10mm of rain over a broad swath extending from Devon and Somerset in the south-west, to Northumberland and Berwickshire in the north-east. In many places it was the driest February since 1821, although it was to be eclipsed just four years later, and farmers in eastern England were already worried about the possibility of a spring drought leading to a poor harvest.

March, April and May passed by with relatively little comment in the newspapers, and in May rainfall reached or even exceeded the average over the bulk of England. By contrast, June brought a combination of drought and heat which considerably exacerbated the situation, drying out fields and meadows and initiating water-supply problems in many parts of the country. Nights were often

Table 8.2 Driest periods of different length during the 1887–88 drought

Period (months)	Months 1887–88	Rainfall (mm)	Percentage	Driest since	Return period (same months)	Return period (all months)
3	Feb.–Apr.	110.5	54	1863	12 yr	3 yr
4	Apr.–July	149.5	61	1870	24 yr	6 yr
5	Feb.–June	185.3	56	1785	48 yr	18 yr
6	Feb.–July	223.4	58	1785	60 yr	26 yr
7	Feb.–Aug.	276.8	60	<1766	120 yr	30 yr
8	Jan.–Aug.	348.6	62	<1766	120 yr	20 yr
9	Feb.–Oct.	442.4	70	1803	48 yr	10 yr
10	Jan.–Oct.	514.2	70	1803	80 yr	10 yr
11	Feb.–Dec.	597.5	70	1854	60 yr	6 yr
12	Feb.–Jan.	636.5	68	1854	80 yr	8 yr
15	Feb.–Apr.	822.4	72	1854	80 yr	8 yr
18	Jan.–June	1003.8	74	1854	60 yr	6 yr

relatively cool, but daytime temperatures were often in the high 20s or low 30s C, and the extended period of hot and sunny weather coincided with the nationwide celebrations marking Queen Victoria's Golden Jubilee. The month's rainfall was little more than 25 per cent of the long-term average over England and Wales as a whole, and closer to 20 per cent in Scotland and in Ireland, while at a more local level large parts of Yorkshire, Lincolnshire, East Anglia and the West Country had less than 10 per cent. What little rain there was fell during the first three days, and most parts of Britain had none at all from the 4th onwards, while several sites in Devon and Cornwall, including Exeter, Liskeard and Bude, reported a completely rainless month. Initially, the fine weather was welcomed by those working the land: 'the wheat crop was immensely improved by the sunshine', but by the end of June farmers were not so optimistic:

'roots and pulses were in want of rain at the close and the water supply was getting short'.

The Manchester correspondent of Symons's *Meteorological Magazine* wrote: 'Rain is much wanted; the reservoirs of waterworks in Lancashire are getting very low, and unless rain comes speedily a scarcity of water will be severely felt in the manufacturing districts.' And from Arncliffe in the Yorkshire Dales: 'Everything suffering from drought; fishes dying from want of water in the beck; and pastures, even [water-] meadows, dried up.' Newspaper reports indicated that water supply was already being restricted in some parts of Lancashire and Yorkshire; for instance, the Burnley Water Company cut off water to industrial concerns in order to preserve a supply to domestic users. G. J. Symons himself talked of 'a general cry of unprecedented drought' although he distanced himself from such hyperbole and indicated time and again with impeccable statistics that it was not so.

July was also sunny, hot and dry – June and July together the hottest for fifteen years, averaged nationally – although rainfall was somewhat higher than it had been during June thanks to occasional scattered thunderstorms. But most of the rain that fell quickly evaporated quickly, and the drought intensified further. Press cuttings reveal a serious state of affairs in many parts of the country, not least in Yorkshire. A report on 4 July quoted farmers in the Doncaster area as saying their crops 'were burnt up and their potatoes spoilt', while the Thirsk district was 'in a fearful state with pumps run dry, and pastures with scarcely a blade of grass'. On 12 July, ponds and wells in upland parts of the country were said to be exhausted, and drinking water had frequently to be fetched long distances. A court instructed one local sanitary authority to reopen a well in Bedale which had been shut down some years previously because of contamination, in order to provide a temporary water supply to local inhabitants. Elsewhere, water was selling at a halfpenny a bucket at Barry, Glamorgan, while in Cumberland 'water is being carried to the grouse'! As July gave way to August, the Revd Charles Griffith wrote from Strathfield Turgiss (now Stratfield Turgis), near Basingstoke: 'The ground is like iron, the skies like brass . . . from being a late season, it has become an early one; wheat, peas, winter oats and barley were being cut at the end of the month, and the country was waiting for rain.'

The drought and heatwave continued until mid-August, but the second half of that month was cooler and less settled with widespread showers. Griffith was again busy with his correspondence: '... vegetation of all kinds suffered, the leaves on many of the trees drooping and withering ... moles died by hundreds, unable to penetrate the iron-bound earth'. The showers did nothing to alleviate the difficulty of supplying water to towns and villages in many parts of England and Wales. During the second week of the month some 4,000 quarrymen in north Wales were thrown out of work when the streams supplying the quarries ran dry; at Llanelli, 1,000 men were similarly laid off at the tin works when the water supply from the Cwmlledi reservoir failed, and not far away at Mountain Ash there was an outbreak of typhoid fever which was attributed to fouled wells. The following week the Rivers Lowman and Exe, at Tiverton and Exeter respectively, were described as so foul that they were little better than sewers. By the end of August the village of Langho, near Blackburn, had been without water for a month; water was being ferried by rail from Clitheroe in milkchurns, and it was doled out by the stationmaster who allowed one bucket per family. The shortage had become so acute in Manchester that the entire city had its water supply cut off at night.

September, mercifully, was a changeable month with abundant rain in western and northern regions, and near-normal rainfall in the Midlands and eastern England. Farmers were not helped, though, by severe gales in the first half of the month and sharp frosts during the second half, both of which further diminished the fruit harvest. Heavy rain during the opening days of September led to a resumption of stream flow in north-east Lancashire, sufficient for many workshops and cotton mills to reopen, and the water level in the Leeds–Liverpool canal rose by 15cm overnight on the 2nd–3rd. In Manchester, however, the situation did not improve, and on the 28th the Corporation's Waterworks Committee heard that the city's stock of water had continued to diminish, and stood at only 26 days. As a consequence the committee voted to retain the restrictions on supplies to both domestic and industrial consumers.

The remaining months of 1887 were rather changeable with frequent falls of rain, although amounts were generally modest, and averaged nationally, the monthly totals in October, November and December were each marginally below the long-term mean. In

Ireland the consequences of the drought were as severe as elsewhere in the United Kingdom, and in County Down the Bessbrook Spinning Company was forced to close its linen mills as the small reservoir and local streams which fed the factory's boilers had all run dry; over 2,000 workers were laid off. And as late as 6 December restrictions in Liverpool were tightened by the Corporation as the water in the Rivington reservoirs reached a dangerously low level.

Rainfall continued generally below average throughout the first half of 1888, as the accompanying tables and graphs clearly illustrate. During the winter months many of the serious local water shortages eased slightly (winter rainfall is usually more than sufficient to offset use, except in the very driest seasons) and the nocturnal severing of supplies in Manchester was ended in March, but during the spring of 1888 reservoir and aquifer levels were markedly lower than they had been at the same season in 1887. Chalk springs in the Downs of Sussex, Surrey and Hampshire, and in the Chilterns north of the Thames, remained dry throughout the winter, and there was therefore much foreboding that another hot and dry summer could result in catastrophic failures in water supply in all parts of the nation. May, in particular, was dry, but the situation eased in June which was an abnormally cool and cloudy month with near-normal rainfall, and the drought effectively ended in July which was one of the coldest and wettest ever known. The farmers now had something different to complain about.

The drought of 2005–6

The drought of 2005–6 had one thing in common with that of 1887–88: there was, once again, a 'general cry of unprecedented drought', although this time there was no one of the stature of G. J. Symons to correct such a misinterpretation of the statistics, although one or two of us did try.

One would scarcely believe it, when confronted with the national print, broadcast and online media headlines, but – unlike 1887 – the year 2005 ranked nowhere in terms of dryness. In fact there had been 21 drier calendar years, averaged over England and Wales, in the period 1905–2004. In a sense that is unfair because, again unlike 1887, the drought of 2005–6 was very regionalized, with a dramatic shortfall in rainfall amounts confined to southern England where, curiously enough, most of the nation's newspapers and broadcasting organizations are based. The London-based news media are conscious of a metropolitan bias in their reporting and do try (some more successfully than others) to counter it, but as far as weather stories are concerned that bias is as strong as ever. Across a narrow zone from Berkshire and Hampshire to west Kent the rainfall during 2005–6 was very nearly as low as it had been in 1887–88, but elsewhere in the UK there was simply no comparison. As one climatologist put it, for much of Britain rainfall was merely 'a bit on the low side'.

Between November 2004 and February 2006, only three of the sixteen months had above-average rainfall, and although spring 2006 was generally quite wet, the drought had a nasty sting in its tail during June and July of that year. Not only were these months very dry, they were also very sunny and very hot – unprecedentedly so in the case of July. It is also important to note that before the dry period began there had been copious rainfall in all parts of the country during August and October 2004 so that ground water levels were well above average at the beginning of November. We may make the same return period calculations as we did with the 1887–88 drought in order to place the drought of 2005–6 in a proper historical context.

The process of reviewing the press cuttings from the 2005–6 drought was very revealing. It seems that no one was seriously affected by it. Not a single person. The acres of newsprint and hours of broadcasting given over to the subject between February 2005 and August 2006 were almost completely concerned with reports of the limited restrictions imposed by certain water companies, speculation about the threat of a water shortage, and the possible repercussions. It was, therefore, a largely political drought characterized by extensive 'spin'.

The dry spell first hit the news towards the end of February 2005, barely three and a half months after it began. The report from the Environment Agency (the EA) noted that reservoir levels in southeast England were lower than normal, and that there could be problems later in the year if the lack of rain continued – sheer speculation at that point. It also spun the statistics by saying that the previous three months had constituted the 'fourth driest winter period since 1883' which sounds much more impressive than the

Table 8.3 Monthly rainfall over England and Wales,
October 2004 to December 2006

Month	Rainfall (mm)	Percentage of 1971–2000 normal
2004		
October	143.1	151
November	51.9	52
December	67.0	66
2005		
January	55.9	60
February	48.0	72
March	57.6	79
April	80.6	129
May	50.1	82
June	62.5	92
July	76.1	133
August	61.7	85
September	71.6	86
October	124.6	131
November	84.7	85
December	73.9	73
2006		
January	30.9	33
February	55.6	84
March	87.1	119
April	53.7	86
May	114.9	187
June	24.9	37
July	37.2	65
August	88.5	122
September	79.2	96
October	111.4	117
November	107.8	108
December	114.6	113

Table 8.4 Driest periods of different lengths during the 2004–6 drought

Period (months)	Months 2004–6	Rainfall (mm)	Percentage	Driest since	Return period (same months)	Return period (all months)
3	Nov.–Jan.	174.8	59	1988–9	20 yr	11 mo
4	Nov.–Feb.	222.8	62	1962–3	33 yr	17 mo
5	Nov.–Mar.	280.4	66	1975–6	33 yr	24 mo
6	Nov.–Apr.	361.0	76	1996–7	10 yr	17 mo
7	Nov.–May.	411.1	77	1975–6	16 yr	24 mo
8	Nov.–June	473.6	79	1975–6	12 yr	30 mo
9	Nov.–July	549.7	85	1995–6	11 yr	24 mo
10	Nov.–Aug.	611.4	85	1988–9	11 yr	30 mo
11	Nov.–Sept.	683.0	85	1995–6	9 yr	30 mo
12	Nov.–Oct.	807.6	89	1995–6	8 yr	13 mo
15	Nov.–Jan.	997.1	84	1995–7	20 yr	17 mo
18	Nov.–Apr.	1193.5	87	1995–7	16 yr	24 mo
21	Nov.–July	1370.5	89	1995–7	16 yr	24 mo

'driest November–January period since 1988–89', or the 'driest three-month period since 2003'. It also quoted an EA spokesman as saying that it was all caused by climate change. No reputable scientist would put his name to such a claim, and mercifully we were treated to very few similar suggestions during the remainder of the drought. But it has to be pointed out that government agencies like the EA, the Met Office and others should be providing the general public with facts, not spin, during episodes such as this.

By the end of March, the headline had changed to the 'fourth driest winter since records began', without specifying when records began, and without specifying whether this referred to the whole of the UK or to a single region. In fact it was, both nationally and for the south-east region, the driest November–March period since 1975–76, and the driest five-month period at any time of the year since 1997. A more objective analysis should also have included October, since the winter gathering period, when reservoirs and aquifers are normally topped up, begins then. October 2004 had

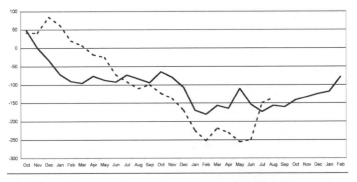

Figure 8.2 Accumulated rainfall deficit (in millimetres) during the droughts of 1886–88 and 2004–7. The graphs begin in October which is the start of the hydrological year when reservoirs and aquifers normally begin their cycle of winter recharge (dashed line to follow 1886–88 and solid line to follow 2004–7)

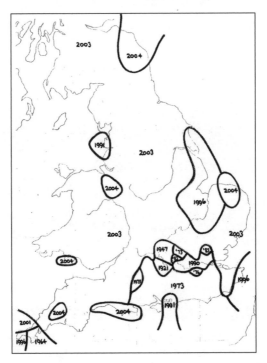

Figure 8.3 Cenoxeric map for 2005: this shows the last year that had lower annual rainfall than 2005

Plate 8.1 Bough Beech Reservoir, Kent, appears to be almost full of water on 15 May 2006, the day the first drought order for ten years was imposed
(Gareth Fuller/PA Wire)

Plate 8.2 Bewl Water, near Lamberhurst, Kent, represents most people's image of a drought in the UK, the water level having fallen to just 30 per cent of normal on 22 February 2006
(Gareth Fuller/PA Wire)

been wet, so the figures immediately become less dramatic. Rainfall averaged over England and Wales from October 2004 to March 2005 inclusive amounted to 423mm, 20 per cent below normal, and this shortfall has a return period of just six years for the specified months, and barely two years at any time of the year. The most recent instance had been in 2003.

The suspicion was that the water industry was talking up the dry spell in order to get domestic users to be more careful in their use of water, just in case – a laudable aim if it had been tackled more honestly. Water companies in general, and Yorkshire Water in particular, had rightly come in for scathing criticism after the over-hyped drought of 1995. It lasted all of five months, the directors of Yorkshire Water resigned en masse after having been forced to ferry water to customers in tankers, and OFWAT gave the water industry a jolly good talking to. It may also be remembered that the ten-month drought in 2003 produced no serious shortages. In other words, the water supply companies appeared to have got their collective act together, and exaggerated warnings of doom seemed to be an important weapon in their armoury in their attempts to reduce profligate consumption. That they gave the impression of targeting domestic consumption rather than commercial and municipal consumption was lost on most commentators.

Between late April 2005 and late February 2006 most of the small water companies operating in south-east England south of the Thames introduced, first, restrictions on sprinklers and unattended hosepipes, and then full hosepipe bans. In July, Folkestone and Dover Water Company asked the government for the right to insist that customers instal water meters. This was initially refused, but the government had performed a U-turn by the following March, thus giving the green light for other water companies to follow Folke-stone and Dover's example at some unspecified time in the future. Richard Aylard, a director of Thames Water, described the spell at the end of June as 'eight months of almost unprecedentedly dry weather': he could get away with it – just – because of that little word 'almost', but of course he knew that everyone reading his comment would see the word 'unprecedentedly' and simply not notice the 'almost'. Using his technique, one might suggest that such a comment is almost a bare-faced lie. Let us look at the statistics again, this time including October 2004 for the reasons noted above.

Total rainfall, averaged over England and Wales, from October 2004 to June 2005 was 617mm, just 11 per cent below the long-term average; there were sixteen drier October–June periods in the previous 100 years. Even in the driest counties of Surrey, Sussex and Hampshire, where the shortfall was 24 per cent, the period ranked only eleventh driest in the last 100 years.

It is illuminating that there appeared to be a surge of news stories during the first few days of each month. The reason was that this was when the previous month's rainfall statistics and reservoir water-level data were released by, respectively, the Met Office and the Environment Agency. All the more reason, then, for these agencies to confine themselves to releasing information rather than comment. During the remainder of 2005 and the early part of 2006, most of the stories were simply reruns of earlier ones, partly because the rainfall statistics became rather less newsworthy as each month passed. The water companies' quite reasonable exhortations not to waste water were, by this stage, accompanied by news releases from

Plate 8.3 Looking west from Dunstable Downs to Ivinghoe Beacon and across Aylesbury Vale, the middle distance is hazy and the countryside is bleached, following 20 months of below-average rainfall, 15 July 2006
(compare Plate 5.4)
(Philip Eden)

other organizations: for instance, the Royal Society for the Protection of Birds warned of a threat to wading birds in river estuaries; the Campaign to Protect Rural England said that the Environment Agency and the water companies had no strategy for dealing with droughts; the Consumer Council for Water called for more investment to prevent leaks, to instal new reservoirs and to build desalination plants; and someone in the Environment Agency was worried about the threat of low water levels to the salmon population on the Hampshire Avon. And finally, at the end of May 2006, the first drought order was put in place for the Sutton and East Surrey Water Company, belatedly extending the restrictions to industrial and other commercial customers, including local authorities, golf clubs, car washes, swimming pools, and so on. Other companies followed suit.

The drought seemed to end in May 2006, which was exceptionally wet from mid-month onwards, and appreciable amounts of rain fell on practically every day from the 17th to the 29th. Averaged over England and Wales the month's rainfall was 90 per cent above normal and the highest in May since 1979. But May's rain quickly evaporated into uselessness during June and July which were not only very dry, but also very hot and very sunny, and for a few weeks it really did seem as though the prospect of rota cuts and standpipes in the streets – threatened for so long – might really come to pass. But the rains did return with a vengeance in mid- and late-August and remained with us throughout the following autumn and winter, finally bringing this very peculiar, very twenty-first-century, drought to an end. Even so, it took a further six months for the hosepipe bans to be lifted: on 18 January 2007 by Thames Water, Southern Water, Sutton and East Surrey Water, and Three Valleys Water; on 7 February by South-east Water, and on 28 February by Mid-Kent Water. And by October 2007 the government announced new regulations to restrict the use of water during a drought, extending those restrictions to hot tubs and jacuzzis, ornamental ponds and fountains, patio cleaners, and private swimming pools. Once again, it seemed, the domestic users were targeted and the commercial ones escaped.

The abiding memories of the drought of 2005–6 were the doomladen warnings of the water industry Cassandras which never quite materialized, and the disgraceful manipulation of the rainfall statistics by organizations which should have known better. As one commentator put it: 'Thirty years ago these spokesmen had the grace to look thoroughly uncomfortable when they lied to us; today they lie with a smile on their face.' Even more to the point, how would the nation cope today with a real drought – one like 1887?

Chapter 9

Compare and contrast: Britain's worst disasters

The British news media love a disaster. And they love all the 'facts' and 'figures' which go with it all. Inverted commas are appropriate because those statements presented as 'facts' are often nothing of the sort, while the 'figures' that seem to add background detail are, even if accurate, routinely provided out of context and may therefore be extremely misleading.

As we have seen in earlier chapters, one of the characteristics of the reporting of weather-related disasters in the press, on radio, on TV and online, is the widespread (perhaps endemic) tendency to exaggerate the rarity of a current or recent event to support an underlying agenda. That agenda may be nothing more than a desire to justify the hyped reporting of the event – a self-fulfilling exercise that is probably most satisfying in its circularity. It is probably not, as some suggest, a conscious attempt to play up the anthropogenic climate-change argument; climate change is sometimes introduced as a likely culprit for the disaster, but that is simply because it adds newsworthiness to the reporting. This, too, becomes self-fulfilling. Expert opinion is rarely sought, either to place the event in a proper statistical and/or historical context, or to examine any relationship between the disaster and climate change, so the errors are repeated and soon – at least in journalistic circles – become conventional wisdom.

A trivial instance of how frightened newspapers now are of accurate but bland journalism will serve to illustrate the point. A recent review of the previous month's weather in one of Britain's most staid broadsheet papers was changed, without reference to the original contributor, from:

November was a generally quiet month: apart from a severe gale in the extreme north of Scotland on the 8th the only windy days were the 18th and 30th . . .

to:

We had it all in November, but apart from a severe gale in the extreme north of Scotland . . .

This enabled the headline writer (presumably the same sub-editor) to add an inappropriately lurid headline.

The mechanism by which the historical importance of a severe-weather episode is exaggerated relies on the fact that, even if we confine our comparison to, say, flooding, the events are all slightly different from each other. The variables include, among others, the speed of onset of the flood, the level of water at different locations, the geographical extent of the flooding, how long it lasts, how many people have to be evacuated from their homes, and so on. The journalist then simply selects whichever of the parameters is most newsworthy. For instance, with reference to the floods of summer 2007, if the water level at Worcester was the highest for a summer flood since 1886, it becomes 'the worst flood since 1886' no matter what the water levels were at Shrewsbury or Bewdley or Tewkesbury or Gloucester, and no matter how many winter floods of greater severity there had been in the interim. Or if the rainfall on one day in June at a single monitoring station in Sheffield was the heaviest in 100 years of records, it becomes 'Yorkshire's worst downpour on record', notwithstanding heavier rain in Leeds in 1982, in Halifax in 1989, in Longdendale in 1973, or in Bradford in 1956.

Government agencies and other information providers frequently assist the sensationalizing of these stories – sometimes unwittingly, sometimes mischievously – by providing statistical background replete with caveats and provisos. These qualifications and exceptions rarely find their way into the media reporting of the event, and never appear in the headlines. Arguably the most insidious of these provisos are the three little words 'one of the . . .' – weasel words if ever there were any. Thus, summer 2007 was reported in government press releases to be 'one of the wettest on record', autumn 2007 was 'one of the driest on record', and a North

Sea storm surge on 9 November 2007 was 'one of the biggest in over two decades'. All good headline material, and so much more impressive than 'the summer ranked 16th in 240 years of records for wetness', or 'the autumn ranked 14th in 240 years of records for dryness', or ' the storm surge was among the ten biggest in the last 25 years'.

Another illustration is relevant here. This is a real exchange between meteorologist and journalist following a localized downpour in south-east London, starting with the journalist's question:

So, was it a record?
No.

Well, then; how much rain was there?
Here at the Weather Centre we had half a millimetre, that's 0.02 of an inch.

Don't be daft; what about the floods? We heard there was something like two inches down in Kent.
There was half a millimetre at the Weather Centre, one millimetre at Heathrow and three millimetres at Gatwick.

That's no use. I can't write a story round that.
Sorry, but they're the facts.

But when was the last time two inches of rain fell in a day at the Weather Centre?
(After a pause, leafing through the record book) **Oh, that was July the 9th in 1981 when 58 millimetres, that's just over two inches, fell in 50 minutes, but you can't . . .**

Yes I can. That's just what I wanted. Bye now.

The front page story appeared under the headline: 'SHOCK FLOODS, 8 MILLION LONDONERS HIT', and described the downpour as 'London's heaviest storm for a quarter of a century'. In fact, fewer than half a million were directly affected, and it was the heaviest downpour anywhere in the London area for precisely two weeks.

The worst disasters

It is a thankless task to attempt to rank individual rainstorms or windstorms or snowstorms in order of severity for the very reasons given above. It is evident that trying to grade disasters of different kinds is to weave a rope out of sand – so foolish, in fact, that we are going to try to do it! The results are achieved by an objective method, and can therefore be reproduced, more or less, by anyone; the scientific rigour, however, is open to challenge. In fact, none is claimed; rather, the challenge is to others to suggest a better method. What we have is, for the first time, an objective ranking of weather-related disasters in the UK which is made available to journalists, politicians and others, so that they have no excuse for failing to put future severe weather events into a reasonably sensible statistical and historical context.

There are many different ways in which disasters may be analysed and many different ways in which they are ranked. These include:

- Number of deaths and injuries.
- Number of people evacuated from their homes.
- Number of people affected, including minor damage, loss of services, etc.
- Damage to the landscape.
- Total financial loss (usually estimated by reference to insurance losses).
- Financial loss as a proportion of the Gross National Product of the affected country.
- The duration of the event.
- The duration of disruption to people's lives.
- The size of geographical area affected.
- The statistical return period of the meteorological extreme.
- The exceedence, geographical and temporal, of selected meteorological thresholds.
- Any combination of the above.

Measuring the human dimension of these occurrences may seem simple at first glance, but collection of the relevant data is fraught with difficulties – omissions, errors, inconsistencies and a general absence of widely accepted definitions. The government departments and agencies responsible for gathering statistics change over time, and there may be as many as three or four sharing that responsibility during any one emergency. Print and broadcast media reporting is heavily weighted towards the day the disaster breaks, when facts and figures are most likely to be wrong, and the journalistic interest declines as each day passes unless there is a renewed threat. The longer-lasting the event is, the more subdued the coverage – droughts constitute a prime example of this. And there is very little retrospective reporting which should allow a more rounded and responsible assessment of the disaster and a more accurate statistical background to be presented.

Even as apparently straightforward a statistic as the death toll associated with a particular bout of severe weather is, in many cases, the subject of controversy. The number of deaths in the UK resulting from the east coast floods of 31 January–1 February 1953, is normally given as 307, but of that total only 155 people drowned in the coastal flooding; 128 others were lost when the ferry *Princess Victoria* sank between Scotland and Northern Ireland; of the remaining 24 deaths some were drowned as several vessels foundered around Britain's coastline, while others died in the severe gales on land. More difficult still is the assessment of the death toll associated with urban smogs: the usual methodology is to calculate the difference between the observed mortality rate and the long-term average at the same time of the year; but when should the cut-off date be, and how can deaths caused by high pollution levels be distinguished from deaths due to, say, severe cold? It has even been mooted in some circles that the lives lost in the great London smog of December 1952 were of less consequence than those lost in other disasters because most of those who died were elderly and chronically ill and 'would have died anyway in a matter of weeks or months'; many readers will find such a view offensive, and the analyses in this book make no judgement about perceived differences in the value of people's lives. Clearly, there are limits to the validity of statistics of this sort, and all the figures published in this work are given with those limitations in mind.

Other aspects of the effect of weather disasters on the population are even more difficult to assign numbers to. Economic losses are typically expressed in terms of insurance losses and government contributions, usually normalized to the value of the pound sterling in a particular year (to eliminate any distortion due to inflation), but these figures actually say more about the size of the population, the wealth of the people affected, the effectiveness of their insurance policies, and indeed the overall take-up of insurance, than they do about the severity of the weather event, and of course all of these aspects vary considerably over time. They also reflect the degree to which the population is exposed to severe weather, and we have seen in an earlier chapter that the acceleration of residential and commercial development on the flood plains of rivers and flood-prone sections of the coastline during the second half of the twentieth century has placed many more people at risk. The insurance industry often cites the rapid increase in losses during recent decades as an indication of climate change. It is nothing of the sort; it is almost entirely a result of the changes outlined in this paragraph.

We should also note that conventional ways of assessing the impact of natural disasters on the nation do not take into account intangible effects such as anxiety, stress and other psychological impacts. Post-traumatic stress is now rather better understood than it was even twenty years ago, and we know that feelings of resentment and anger, blame and guilt can persist for years.

With all these caveats in mind, we can now look at major weather catastrophes of past times, and try to compare one with another. Thanks to the substantial number of publications, especially periodicals, between approximately 1865 and 1990, and the often extensive and meticulous accounts of newsworthy weather events found therein, this period provides the most comprehensive data on weather-related disasters. These years also coincide with diligent and detailed reporting in the UK's broadsheet newspapers. Since 1990 the number of publications of record in the meteorological/climatological sector has declined almost to zero, and the analysis of extreme weather events by government agencies has also fallen sharply; these accounts are now largely provided in other periodicals by interested and knowledgeable individuals. Since 1990, too, the detail and reliability of newspaper reporting have declined, while the more recent explosive growth of online accounts, while welcome, is extremely variable in quality and contains many gross errors. Before 1865, there is a variety of sources containing descriptions and data for many extreme episodes, but a substantial number of medium-sized and smaller events have probably been missed, and the scientific content in the reports diminishes quickly the further back in time one goes. Nevertheless, we know a good deal about Britain's most dramatic natural disasters all the way back to the eleventh century.

Windstorms

Professor Hubert Lamb, in his book *Historic Storms of the North Sea, British Isles and North-west Europe*, makes a rare attempt at grading windstorms in his area of interest, which includes a much wider geographical region than just the British Isles or the UK. His Storm Severity Index, a measure of the intensity of the meteorological phenomenon rather than an assessment of its impact on the human population, takes the form:

Wind power × area affected by damaging winds × duration of damaging winds

where wind power is represented by the cube of the maximum gradient wind speed. Using this definition he found that the most severe storm was that of 15 December 1986, the occasion of the deepest Atlantic depression on record which was centred just off the southern Icelandic coast on that date; this depression brought very disturbed weather with heavy rain and gales to the UK, but it was responsible for relatively little damage or disruption. This was followed in the rankings by the storms of 10–12 December 1792, 4 February 1825, 31 October to 2 November 1694, and 7–8 December 1703 (all dates prior to the calendar change in 1752 are converted to the present Gregorian calendar). Those of 1792 and 1825 scored highly because of their long duration, while that of 1703 is arguably the most infamous gale of the millennium, having occasioned widespread destruction and loss of life. Lamb's Storm Severity Index, then, does not coincide with our normal understanding of a 'disaster' which is heavily inclined towards loss of life, material losses and severe disruption to normal activity, but it provides an important additional dimension to our knowledge of extreme meteorological events. Above all, it emphasizes that disaster analysis is a flawed and very inadequate surrogate for any assessment of climate change.

Table 9.1 Most extreme windstorms over the British Isles using different parameters

Meteorological severity	Loss of life (adjusted to a national population of 60 million)	Insurance losses (normalized to 2008 prices, since 1900 only)
1703, 7–8 Dec.	1703, 7–8 Dec.	1987, 15–16 Oct.
1839, 6–7 Jan.	1588, 21 Sept.	1990, 26 Jan.
1953, 31 Jan.–1 Feb.	1695, 22 Sept.	1962, 16–17 Feb.
1990, 26 Jan.	1953, 31 Jan.–1 Feb.	1976, 2–3 Jan.
1976, 2–3 Jan.	1879, 28 Dec.	1903, 26–27 Feb.
1886, 8–9 Dec.	1839, 6–7 Jan.	1927, 28 Jan.
1792, 10–12 Dec.	1990, 26 Jan.	1997, 24–25 Dec.
1884, 25–27 Jan.	1886, 8–9 Dec.	1961, 16–17 Sept.
1893, 17–19 Nov.	1987, 15–16 Oct.	1968, 14–15 Jan.
1987, 15–16 Oct.	1791, 1–2 Mar.	1953, 31 Jan.–1 Feb.

Some of these gales, notably 31 January 1953, were also associated with widespread flooding which brought extensive damage and loss of life on its own account; where possible only the losses due to the strength of the wind (including those in home waters) have been included here.

Coastal floods

Although associated with powerful north-westerly and northerly gales, the North Sea storm surges and the resulting coastal flooding along the east coast of England are quite distinct phenomena, depending also upon the magnitude and timing of the astronomical tides. Storm surges are also prevalent along other parts of the British coastline, notably in the Bristol Channel, along the north Wales and Lancashire coasts, in the Solway Firth, and in some Scottish sea lochs. Coastal flooding may also result from powerful wind-waves and long-period swell: such events are relatively common along the English Channel coast.

Table 9.2 Most extreme coastal flooding events over the British Isles using different parameters

Meteorological severity	Loss of life (adjusted to a national population of 60 million)	Insurance losses (normalized to 2007 prices, since 1900 only)
1953, 31 Jan.–1 Feb.	1953, 31 Jan.–1 Feb.	1953, 31 Jan.–1 Feb.
1978, 11–12 Jan.	1881, 18 Jan.	1928, 6–7 Jan.
1881, 18 Jan.	1928, 6–7 Jan.	1881, 18 Jan.
1928, 6–7 Jan.	2005, 11–12 Jan.	1978, 11–12 Jan.
1990, 26 Feb.	1978, 11–12 Jan.	1990, 26 Feb.
1983, 1 Feb.		1988, 3–4 Mar.
1977, 11–12 Nov.		1979, 13–14 Feb.
1988, 3–4 Mar.		1998, 4–7 Mar.
2005, 11–12 Jan.		1996, 19–20Feb.
1979, 13–14 Feb.		1983, 1 Feb.

River floods

Repeated flooding in different parts of Britain characterized the summer of 2007, and brought home to many that urban development on flood plains simply places more home-owners and more businesses at risk. In the past decade or so the waters have, so to speak, been muddied by the introduction of climate change into the discussion of whether it is sensible, or economically viable, to build in flood-prone localities. Recent rainstorms have all fallen within the boundaries of previous observational experience, so it is not necessary to invoke climate change as a possible culprit in what appears to be a growing intensity and frequency of recent river floods. Quite the contrary, in fact: it demonstrates the impact human activity has had on the severity of flooding along the courses of Britain's major rivers *before* any systematic change to the frequency of severe rainstorms – predicted by the computer models – has come into play. The thaw floods of March 1947 are included here.

Table 9.3 Most extreme coastal flooding events over the British Isles using different parameters

Meteorological severity	Loss of life (adjusted to a national population of 60 million)	Insurance losses (normalized to 2007 prices, since 1900 only)
1947, Mar.	2000–1, Oct.–Feb.	2000–1, Oct.–Feb.
2000–1, Oct. to Feb.	1960, Oct.–Dec.	2007, June–July
1960, Oct. to Dec.	1994, Dec.	1968, Sept.
1968, Sept.	1829, Aug.	1947, Mar.
1912, Aug.	2007, June–July	1994, Dec.
2007, June–July		1998, Oct.–Nov.
1894, Nov.		1960, Oct.–Dec.
1948, Aug.		1968, July
1829, Aug.		1912, Aug.
2007, June–July		1990, Feb.

Flash floods

By their very nature, flash floods affect very small parts of the country, and have usually come and gone in a matter of hours. But the rapidity of onset of these phenomena – their suddenness – means that it is impossible to prepare for them, and very difficult to save lives and property once they are raging.

Snowstorms

Individual snowstorms are among the best documented meteorological events of the last two centuries. Even so, it sometimes difficult to tease apart the effects on the community of heavy snow, high winds and rough seas, especially for storms which occurred before the twentieth century. We observed in earlier chapters that the disruption caused by relatively small amounts of snow has grown sharply in recent decades as we have become, at one and the same time, a much more mobile and a much more congested nation.

Table 9.4 Most extreme flash floods in the British Isles using different parameters

Meteorological severity	Loss of life (adjusted to a national population of 60 million)	Insurance losses (normalized to 2007 prices, since 1900 only)
1955, 18 July, Martinstown	1952, Aug.	1952, Aug.
1917, 28 June, Bruton	1900, Aug.	2004, Aug.
1924, 18 Aug., Cannington	1920, Oct.	1975, Aug.
1952, 15 Aug., Lynmouth	1900, July	1989, May
1957, 8 June, Camelford	1960, Oct.	1900, July
2004, 16 Aug., Boscastle		
1989, 23 May, Halifax		
1956, Aug., Bradford		
1960, Oct., Horncastle		
1920, May, Louth		

Table 9.5 Most extreme snowstorms in the British Isles using different parameters

Meteorological severity	Loss of life (adjusted to a national population of 60 million)	Insurance losses (normalized to 2007 prices, since 1900 only)
1881, 18–20 Jan.	1891, 8–10 Mar.	1891, 8–10 Mar.
1891, 8–10 Mar.	1881, 18–20 Jan.	1881, 18–20 Jan.
1927, 25–26 Dec.	1836, Jan.	1947, 5–6 Mar.
1947, 5–6 Mar.	1814, Jan.	1987, 12–14 Jan.
1814, Jan.	1978, 27–29 Jan.	1963, 26–29 Dec.
1978, 18–20 Feb.		
1933, 23–26 Feb.		
1940, 26–28 Jan.		
1962, 26–29 Dec.		
1978, 27–29 Jan.		

Ice storms

Fortunately infrequent in the British Isles, an ice storm is the extreme example of the phenomenon that we should normally describe as freezing rain. Because of the very precise meteorological factors required for rain to fall from a layer of relatively warm air aloft, through a layer of colder air near the ground, and then to freeze on contact, these events are normally very localized and also very short lived. The archives, however, do reveal one ice storm that was severe, widespread and long lasting. It was not reported at the time, as it occurred during the first year of the Second World War, and reports of a Britain largely paralysed by the weather would have been considered useful to the enemy.

The UK's most noteworthy and disruptive freezing rain events in the last 100 years were:

1940, 26–30 January, a broad zone extending from Sussex and
    Hampshire to north Wales.
1966, 20 January, south-east England, Wessex, Thames Valley, south
    Midlands.
1979, 23–24 January, south and south-east England, south Midlands.
1927, 21 December, southern England, including London.
1947, 3–4 March, south-east England.
1963, 3–4 January and 8 February, southern England, Midlands.
1968, 24 December, Wales.
1995, 30 December, southern England and south Wales.
1969, March, upland parts of northern England.

Severe winters

It is true that snowstorms – and the occasional ice storm – are responsible for many deaths and serious injuries and much disruption to people's day-to-day lives, but the additional effects of a severe winter are also serious, although they may be much more insidious. Hypothermia kills many, especially but not exclusively among the elderly, and the number of road accidents increases. Damage to roadways and footpaths caused by frost heave, damage to homes resulting from burst pipes, and disruption to commerce due to power failures, all contribute significantly to the difficulties presented by such a prolonged period of adverse weather. Agriculture is always badly hit in such a winter, and one must not forget either the effects on wildlife.

Thanks to the work of the late Professor Gordon Manley we have a very good idea of the behaviour of temperature in central England over the last three and a half centuries, and the severest winters during that period were, with the Central England temperature:

1683–84	–1.2°C	1694–95	0.7°C
1739–40	–0.4°C	1878–79	0.7°C
1962–63	–0.3°C	1715–16	0.8°C
1813–14	0.4°C	1697–98	1.0°C
1794–95	0.5°C	1946–47	1.1°C

Droughts

Droughts are the most slow acting of all potential weather-related disasters. It takes many months, perhaps more than a year, for our attitude to shift from an initial pleasure that the incessant rains of earlier seasons have stopped, through a slight but growing unease that if the dry spell continues there may be a water shortage some time in the future, eventually to a final realization that disruption of the water supply is imminent.

It is relatively easy to derive a Drought Severity Index (DSI) based on the shortfall in the quantity of rain, the geographical extent of the shortage, and also its duration. A simple way of doing this is to calculate the length of time rainfall remains below 80 per cent of the long-term average, that 20 per cent shortfall representing the threshold beyond which there are likely to be serious and widespread repercussions to the nation's water supply. This figure is then combined with the maximum twelve-month rainfall shortfall to produce an index which takes into account both intensity and duration.

Using this calculation, the worst droughts in the last 250 years over England and Wales as a whole, where the DSI equals the duration in months multiplied by the maximum twelve-month rainfall shortfall expressed as a percentage, were as shown in Table 9.6.

The drought of 1887–88, featured in the chapter on droughts, ranked only twelfth as a consequence of its relative shortness, while

Table 9.6 Worst droughts in the past 250 years

	Duration	Max shortfall	DSI
1783–85	30 months	43%	1290
1853–55	31 months	34%	1054
1975–77	25 months	39%	975
1780–82	28 months	34%	952
1801–03	29 months	32%	928
1920–22	26 months	34%	904
1995–97	30 months	27%	810
1932–34	26 months	31%	806
1843–45	26 months	31%	806

the drought of 2004–6 scored zero, emphasizing that it was severe only in a very small geographical area.

Heatwaves

It is really only in the past decade that heatwaves have been acknowledged as a serious threat to the health of the nation, although there were a number of heat-related deaths during the long periods of very hot weather that occurred during the summer of 1976. In 2003, some 30,000 people died across Europe in a series of unprecedented heatwaves, the majority of them elderly, and in France the national death toll of almost 15,000 had serious and long-lasting political repercussions. In the UK the official figures indicated that over 900 heat-related deaths occurred during that August, although the mortality rate was 2,040 above normal.

In earlier times, hot summers brought widespread disease, including dysentery epidemics, and there were sometimes thousands of deaths. Seriously inadequate arrangements for the disposal of sewage, poor hygiene among the more deprived classes, and periodic failures of local water supply exacerbated the situation.

The hottest months in the Central England temperature record are:

2007	July	19.9°C
1983	July	19.5°C
1995	August	19.2°C
1997	August	18.9°C
1783	July	18.8°C
1975	August	18.7°C
1852	July	18.7°C
1976	July	18.7°C
1947	August	18.6°C
1921	July	18.5°C

Other months with extended heatwaves, partially offset by brief cooler interludes, were June 1846, July 1859, July 1868, July and August 1911, June 1976, and August 2003.

Smogs

We have seen that the incidence of smoke-fogs in Britain's conurbations has declined almost to zero since the introduction of clean air legislation from the 1950s onwards. Before that time, however, the elevated death rates during and after periods of dense and persistent smog indicate that this particular kind of weather probably killed more people in the UK than any other. The most serious smoke-fogs in London in the last two centuries occurred in December–January 1813–14, December 1873, December 1886, January 1887, December 1889, December 1944, November 1948, December 1952 and December 1962. Had the capital's atmosphere not been cleaned up there would certainly have been further serious smog episodes in December 1975, December 1990, and of course December 2006.

During the twentieth century there were also serious smogs in central Scotland in December 1935, in the Lancashire and Yorkshire conurbations in November 1936, and in many parts of the country during March 1953.

Thunderstorms, hailstorms and tornadoes

These meteorological phenomena generally affect very localized areas, and loss of life rarely amounts to more than isolated ones and twos. However, the tornado which struck Edwardsville in

Glamorgan on 27 October 1913 killed six people, while seventeen people lost their lives in and around London during the 'Derby Day storms' on 31 May 1911. The most destructive hailstorms were arguably those that affected various parts of London and the Home Counties around 24 June 1897.

Although mercifully few people have died during tornadoes in Britain, these extremely violent but very localized storms sometimes cause a considerable amount of damage, especially when they strike urban areas. Among the most costly British tornadoes of the last 100 years were the Birmingham tornadoes of June 1931 and July 2005, the west London tornado of December 1954, the long-lasting Buckinghamshire/Bedfordshire/Cambridgeshire tornado of May 1950, and the tornado swarms which affected widely separated districts in November 1981 and December 1990.

What were Britain's worst weather disasters?

In an attempt to answer that impossible question, a disaster index has been devised which takes into account the death toll, the number of people whose lives have been seriously disrupted over a long period, and the meteorological severity and rarity of the event. The index provides a measure of objectivity in ranking such widely differing phenomena as snowstorms, heatwaves, floods and fogs, but the design of the index is itself partially subjective. It will therefore, no doubt, trigger criticism from a variety of sources, but it does try to answer that ubiquitous question, and it is up to others to improve upon it. The results are as follows:

The great gale of 7–8 December 1703.
The London smog of December 1952.
The gale and North Sea floods of 31 January–1 February 1953.
The snowstorm and London flood of 18–20 January 1881.
The so-called 'Burns' Day' gale of 26 January 1990.
The West Country snowstorm of 8–9 March 1891.
The thaw floods of March 1947.
The paralysing snowstorm and ice storm of 26–30 January 1940.
The long-lasting and widespread autumn/winter floods of
    2000–1.
The gale of 15 October 1987.

Chapter 10

... and the next disaster, please

From the global . . .

'How can you predict what the climate will be like in a hundred years time when you can't even get tomorrow's forecast right?' That is the usual response from the layman to the climatologist expounding the latest theory on climate change.

The answer is that the climatologist and the weather forecaster are looking at the behaviour of the atmosphere on two completely different levels. Put it this way. Forecasting tomorrow's weather is a bit like estimating how much loose change you will have in your pocket or purse in 24 hours' time. It is the result of many small transactions, often interrelated, most of them entirely predictable at such short range: a visit to the cashpoint, buying groceries, pocket money for the kids, and so on. By contrast, foreshadowing changes in the climate over a long period perhaps corresponds to calculating the household budget over a year or more: the daily transactions hardly matter, whereas much more important are outside influences, many of which are predictable but some of which may be quite unexpected.

Thus climatologists believe they understand many, but not all, of these 'forcing factors' and are therefore able to make broad-brush, qualified assessments of where the world's climate may go during coming decades.

We can get a feel for the direction the climate is taking by looking back 50 years – not anecdotally, but statistically. During that period the mean temperature of the lowest layer of the atmosphere, where human beings live, has risen by approximately 1°C. The warming has not been even: the northern hemisphere has warmed more than the southern hemisphere, the continents more than the oceans, the polar fringes more than the core of the Arctic and Antarctic, and Europe more than North America. In Europe (excluding the Mediterranean Basin) there were twice as many heat-related deaths in two weeks in August 2003 than there had been in the twenty previous years put together. In September 2005 both the geographical extent and the mean thickness of Arctic Ocean ice reached record low levels, but these records were in turn smashed out of sight in September 2007.

There are signs that the warming process is accelerating. Computer models indicate a rise in temperature, averaged globally, of between 1.5 and 6°C between the years 2000 and 2100. We can expect the hemispheric and continent/ocean differentials to continue, though not necessarily the transatlantic one, so substantial further warming is likely over both Europe and North America.

In the Arctic Basin sea ice may vanish altogether in the late summer before 2020; this will be the first large-scale (non-atmospheric) environmental change triggered by the changing climate, and it will probably have a positive feedback effect. For the time being, energy in the form of latent heat is absorbed throughout the summer by the melting process in the Arctic, maintaining cold conditions there and in particular preventing the ocean temperature from climbing more than a degree or so above zero. If all the ice were routinely to disappear by, say, late July, that energy absorption would halt, the Arctic Ocean would warm by several degrees during the remainder of the melting season – that is, until the end of September – and that in turn would delay the onset of ice formation in the autumn. The change in temperature distribution in the Arctic would also affect ocean currents in the Atlantic which would in turn influence the atmospheric circulation in the region.

These knock-on effects are very difficult to model on the computer because we have no detailed measurements from previous such occurrences: they are, if you like, the climatological equivalent of the impact that a demotion at work or an unexpectedly large tax bill would have on the family budget. However, one might postulate a poleward shift in the Atlantic depression track, and that would leave much of Europe – north-western Britain, Iceland and Norway excepted – with less rain in all seasons and therefore more prone to water shortages.

Tropical revolving storms have, ever since the 1980s, been regarded as particularly sensitive to a changing global climate. Known as hurricanes over the Atlantic and north-eastern Pacific, typhoons in the north-western Pacific, and cyclones in the south-western Pacific and Indian Ocean, these storms are known to develop regularly only where the sea-surface temperature exceeds 26°C. Although their frequency has increased sharply in the last fifteen years in the Atlantic/Caribbean sector, most climate experts regard this as part of a natural 60-year cycle – there were previous peaks in the 1940s and 1880s – and the trigger for the formation of a hurricane is meteorological rather than climatological. Nevertheless, a warming climate will probably result in storms which are longer lasting and more intense, and which may develop in areas hitherto largely immune – offshore Brazil, for instance. And although their frequency may decrease between now and about 2030 in line with the natural cycle, this may be partly offset by an extension of the season both in spring and autumn.

Remember, too, that the coastal fringes threatened by these storms, from Texas to Taiwan, Florida to the Philippines, are becoming increasingly urbanized and therefore increasingly susceptible to huge human and economic losses. The British Isles has an interest in hurricanes too – many of the most severe and destructive autumn gales experienced in our part of the world are associated with mid-latitude depressions which contain the remnants of defunct hurricanes and tropical storms.

Climate, it was once said, is average weather. That is not so. The climate of a given place is described by the extremes as well as the averages. Even if the world's climate were static there would be dozens of natural disasters every year. With a rapidly changing climate, the next 50 years may turn out to be a white-knuckle ride: droughts, floods, heatwaves, hurricanes, even if they do not occur more frequently, are likely to affect wider geographical areas including regions previously untouched, and are also likely to occur during seasons previously exempt. Even the mundane will change; in 2100 the ordinary day-to-day weather where you live will be different, and everything that depends on the weather – flora, fauna, agricultural practices, flood defences, house-building styles, health-service procedures during hot weather, and so on, will have to change too.

. . . to the parochial

Extreme value analysis is a statistical tool which can be very useful in estimating the frequency of recurrence of exceptional events, but the technique has to be used with caution. It was extensively used in climatological studies for a large part of the twentieth century when conventional wisdom was that the climate, although changing very gradually over long periods of time, could be considered to be static for statistical purposes. Thus it could be said that, for instance, the North Sea storm surge of 31 January 1953 had a return period of 200 years, or that a winter as severe as that of 1962–63 could be expected to recur, on average, once every 250 years. Such statements should also be qualified by quoting the 'error bars', and the error bars become very wide when the calculated return period is longer than the climatological record from which come the data used in the analysis. For example, if 200 years of temperature records are used to examine cold winters, and the statistical technique tells us that the return period of a particular severe winter is 250 years, it will also tell us that the error bars, depending on the kind of distribution, may be plus or minus 150 years. In other words, the event may recur, on average, between once every 100 years and once every 400 years.

The accelerating rise in temperature, observed both in the British Isles and globally since the 1980s, means that all calculations based on an essentially static climate go out of the window. This is particularly and most self-evidently true of the frequency of extreme temperatures – daily, monthly, annual – but it is probably also true of other climatic parameters, especially those closely dependent on temperature such as the frequency and severity of frost, the number of days with snow falling, and the number of days with snow lying. Even those elements not closely related to temperature, such as rainfall and wind speed, will be affected because changes in the temperature of the air alter the dynamic equilibrium of the atmosphere, and upon that depend all aspects of the climate.

It is fashionable in journalistic and political circles in the UK to ascribe all exceptional weather events to the present warming trend in our climate. Another fashionable view is that weather extremes will occur more frequently and become even more extreme in the future – in fact it is often said that this is already happening. These two views, naturally, feed off each other. But, with a solitary exception, there is no systematic evidence that either is true – not yet, at any rate.

That one exception is the increase in frequency and intensity of summer heatwaves. This is the most direct response to the underlying warming trend in Britain's climate, and it is offset by a decline in the frequency and intensity of severe weather in winter. Thus the sum total of extremes has, it can be argued, not changed significantly. Every other weather extreme experienced in the British Isles in the last two decades has fallen within the bounds of previous experience, as the preceding chapters have demonstrated.

The purpose of this argument is not to make light of the threat of climate change. It is first and foremost to set the record straight, but, just as important, it is to emphasize that we seem, for a variety of reasons, less and less able to cope with relatively routine severe weather episodes. Furthermore, politicians and journalists use climate change as an excuse for our failures. If we are so poorly equipped to respond to disasters which have occurred so regularly in past decades and centuries, how will we fare when we are faced with events outside that envelope of previous experience? Even if, in aggregate, the frequency and intensity of weather disasters do not change, and some climatologists argue that they may not, the make-up of that aggregate will certainly change as the climate itself changes. Put at its crudest – and cruellest – fewer old people will die from hypothermia in winter, but more will die from dehydration and heat exhaustion in summer.

Let us now examine the changes to the character of weather extremes in the UK which might occur in the coming decades. Assuming a continued warming trend, we can, perhaps, identify three types of change, categorized according to how directly a consistently rising temperature might affect them: first, the immediate effect of higher temperatures; second, the effect of higher ocean temperatures in home waters, less ice in the Arctic, and a reduced winter snow-cover over Europe and Asia; and third, the effect of all these changes on the nature of synoptic features – travelling weather systems – which affect north-west Europe through the course of a typical year, and which control the character of our day-to-day weather.

Rising temperatures

We have already noted that the most direct effects of an underlying warming trend are increases in the frequency and intensity of summer heatwaves. The UK national extreme temperature record which had stood for most of the twentieth century was broken in 1990 and again in 2003. The succession was:

36.7°C at Raunds, Northamptonshire, on 9 August 1911.
37.1°C at Cheltenham, Gloucestershire, on 3 August 1990.
38.1°C at Kew Gardens, London, on 10 August 2003.

In a stable climate with long-standing records one would expect the national extreme to be broken rarely, and then only by incremental amounts. The 1990 event just about fulfils these criteria, but the 2003 event breaks a recent record by a wide margin. (The sometimes quoted 38.5°C at Faversham, Kent, in August 2003 is now widely considered to be erroneous.)

Maxima of 35°C or higher have been observed in the British Isles (excluding the Channel Islands) in 1876, 1881, 1900, 1906, 1911, 1923, 1932, 1948, 1957, 1976, 1990, 1995, 2003 and 2006. This amounts to approximately once per decade until 1990, and then four times in the next seventeen years. Maxima of 33°C or higher were reported in only two years in the 1960s, two years in the 1970s, again two years in the 1980s, five years in the 1990s, and in five of the first eight years of the 2000s. And all this comes courtesy of a rise in the overall mean temperature for the summer quarter in central England since 1950 of just 1°C.

Computer models indicate a probable further warming by the end of the twenty-first century in the range 1.5 to 6°C. Taking a relatively conservative estimate of a 3°C rise by the year 2100, what kind of heatwaves might we expect during the next 100 years? One way of answering this question is to look at the twentieth century's most remarkable summer, 1976, and consider what would happen if the synoptic events – the weather patterns – during that season were to be repeated in, say, London, Manchester and Glasgow (Table 10.1). We are making the assumption that the overall temperature rise is identical for all synoptic types, and that is actually a big assumption, but it provides as good an approximation as any.

Table 10.1 Projected temperature rises 1976–2100

Year	1976	2035	2100
Central London (St James's Park)			
June average max temp	25.6°C	27.6°C	29.6°C
June average min temp	14.0	16.0	18.0
July average max temp	26.5	28.5	30.5
July average min temp	15.6	17.6	19.6
August average max temp	24.8	26.8	28.8
August average min temp	14.3	16.3	18.3
No. of days over 25°C	51	67	81
No. of days over 30°C	18	24	46
No. of days over 35°C	0	6	14
Longest run of days over 30°C	7	16	18
Central Manchester (old weather centre)			
June average max temp	21.6°C	23.6°C	25.6°C
June average min temp	13.5	15.5	17.5
July average max temp	23.2	25.2	27.2
July average min temp	14.8	16.8	18.8
August average max temp	24.4	26.4	28.4
August average min temp	14.2	16.2	18.2
No. of days over 25°C	33	45	53
No. of days over 30°C	9	19	27
No. of days over 35°C	0	0	4
Longest run of days over 30°C	9	12	14
Central Glasgow (old weather centre)			
June average max temp	18.7°C	20.7°C	22.7°C
June average min temp	12.1	14.1	16.1
July average max temp	21.5	23.5	25.5
July average min temp	13.8	15.8	17.8
August average max temp	21.9	23.9	25.9
August average min temp	12.4	14.4	16.4
No. of days over 25°C	20	27	31
No. of days over 30°C	0	5	16
No. of days over 35°C	0	0	0
Longest run of days over 30°C	0	3	8

The probability of a recurrence of a summer as hot as 1976 was estimated at the time as once every 200 years, on average, with widely drawn error bars. By the middle of the twenty-first century, if the predicted warming is spread evenly throughout the year, similarly hot summers can be expected to occur once every two or three years. By the end of the century, nine out of ten years are likely to be as hot as 1976 or hotter, and in the most extreme heatwaves by 2100 the temperature will probably exceed the 40°C mark in the nation's hottest spots, and could well climb as high as 41°C. Of course, the degree of warming in the next hundred years may be less than we have allowed for, but it could be more.

We can carry out a similar exercise for severe winters by seeing what would happen to temperatures if the synoptic patterns which produced the infamous winter of 1962–63 were to be repeated. Rather than calculate notional figures for three different parts of the country, let us assess what the coldest winters of the future might be like by examining one location, Edgbaston Observatory in Birmingham, which was in the coldest part of the country in 1962–63 (Table 10.2).

Table 10.2 Projected changes in temperature and in the frequency of frost and snow, 1962–2100

	1962–63	*2035*	*2100*
Edgbaston Observatory			
December average max. temp.	4.0°C	6.0°C	8.0°C
December average min. temp.	0.1	2.1	4.1
January average max. temp.	−0.2	1.8	3.8
January average min. temp.	−3.7	−1.7	0.3
February average max. temp.	1.0	3.0	5.0
February average min. temp.	−3.1	−1.1	0.9
No. of air frosts	74	48	24
No. of days below 5°C	78	65	39
No. of days below 0°C	34	18	3
Longest run of days below 0°C	10	5	2
No of days with snow cover	69	31	7

Cold and snowy winters will very soon be a thing of the past, if they are not already. The winter of 1962–63 was the coldest for over two centuries – since 1739–40 to be precise – and the probability of a similarly cold one occurring now is vanishingly small. A more recent cold winter such as that of 1995–96 may be a more useful point of reference. That particular season was not exceptional, apart from Christmas week in Scotland and a notable February snow-storm in north-west England and south-west Scotland, and few people in eastern, central and southern England will remember much about it. Snow lay on the ground over the English Midlands for between 15 and 25 days, and in southern counties of England for fewer than ten days. It ranked 14th coldest of the last century; put another way, a winter of similar severity would be expected to recur roughly once every seven years, on average. By 2035, such a winter would have a return period of about 33 years, and by 2100 the return period would have grown to over 100 years. So even a cold winter as modest as that of 1995–96 would have all but vanished by the end of the present century. Even with a 3°C rise in the average temperature of the British winter, snowfalls are not ruled out by any means, but they will have become quite rare, and they will probably have the same sort of rarity and novelty value that snowfalls in Algiers or Lisbon or San Francisco have today. No doubt when they do occur they will continue to cause serious disruption to people's day-to-day lives. Again, it should be emphasized that the degree of warming in our winter weather may be less, but it may also be more, than has been allowed for in this analysis.

The changing environment

A continued warming will not only affect air temperatures directly, it will also change the character of the UK's climatological hinter-land: warmer oceans, much warmer waters in the shallow seas around our coasts, less ground moisture over the European land-mass, a rapid diminution of Arctic ice especially in summer and autumn, less snow and ice over Eurasia in winter and spring, and the possibility of cold, fresh meltwater entering the northern periphery of the Atlantic Ocean from the melting Greenland ice-cap.

The present generation of climate-prediction computer models is the first to attempt to foreshadow these sorts of changes, and their results are bound to be simplistic. Their main value is that future generations of computer models will learn from them. For the time being, though, their results tell us no more – possibly less – than informed speculation by experts, and sometimes some climatologists need reminding of that before they open their mouths to brief politicians and journalists. With that huge caveat, let us speculate.

A disproportionate warming of shallow coastal waters is likely to be most apparent during the summer, particularly strongly marked in the inter-tidal zone where the general rise in ocean temperature will be amplified by the warming effect of sun-heated sand, mud and rocks on an incoming tide. Further sea-surface warming may occur if there is, as some models suggest, an increase in frequency of anticyclonic weather types during the summer half-year, especially in southern Britain: such a scenario would be accompanied by a rise in the number of sunshine hours and a reduction in the strength of the wind (and therefore a reduced level of turbulent mixing in the surface layer of the sea). The impact of this offshore warming would be most seriously felt around the coastal fringes of the country during prolonged heatwaves, and it would manifest itself in higher humidity levels and higher night-time temperatures, both of which would threaten public health, particularly of the very young and the very old. As a very rough guide, the minimum temperature in towns and cities around our coastline during extended heatwaves approximates to the temperature of our coastal waters. Already, during the hot weather of August 2003 and July 2006, we have observed sea temperatures of 21–23°C close to the coastline, with overnight lows of 23–24°C along the English Channel coast and in the Channel Islands on the warmest nights during those summers. Given that these shallow waters may warm disproportionately, we may expect temperatures in southern coastal districts of the UK to remain above 27 or 28°C during the warmest nights by the end of the present century.

Taking a broader view, we would expect surface water temperatures in all sea areas around the British Isles, including the deep oceanic basins of the Norwegian Sea and the North Atlantic Ocean, to climb in line with the general planetary warming. Thus, in a hundred years time, we might expect late winter sea-surface temperatures to range from 8°C in the southern North Sea to 13°C in the Southwest Approaches (the mid-twentieth century equivalents were 4°C and 9°C), while late summer values might range from 17°C off northern Scotland to 21°C in the English Channel (mid-twentieth-century equivalents were 13°C and 17°C). This is likely to have wide-ranging repercussions on all aspects of the British climate because nearly all our weather is imported; to reach us it has to cross the various bodies of water which surround us; and as these 'foreign' air masses travel across the water – even the relatively narrow English Channel or southern North Sea – they are appreciably modified in their lowest layers. In respect of disaster generation, however, we may restrict our speculation to those slow-moving shallow summer depressions which are responsible for long-lasting downpours and consequent flash floods in the West Country.

It was noted in an earlier chapter than the majority of daily rainfall totals in excess of 200mm in the UK occur during the summer months in south-west England:

28 June 1917	242.8mm	Bruton, Somerset
18 August 1924	238.8mm	Cannington, Somerset
15 August 1952	228.6mm	Longstone Barrow, Devon
18 July 1955	279.4mm	Martinstown, Dorset
8 June 1957	203.0mm	Camelford, Cornwall
16 August 2004	200.4mm	Otterham, Cornwall

There are also many other examples of falls between 150 and 200mm which were also responsible for serious flooding. Indeed, the short, steep valleys of north Cornwall and north Devon are particularly vulnerable to localized summer downpours. They collect water efficiently from the surrounding moors, channel it rapidly into the main stream, and take it all out to sea in a matter of four to six hours. Because of their almost instantaneous hydrological response to a sudden cloudburst, these valleys are known in the trade as 'flashy catchments' and they produce true 'flash floods'. In the rush to find a scapegoat in the wake of the Boscastle flood in August 2004 – climate change being the favoured one – most commentators ignored the fact that we were long overdue one of these catastrophes, the most recent previous flash flood in south-west England having occurred at Helston, south Cornwall, in June 1993. The 1950s and 1960s gave us several major West Country floods, including the Wadebridge and Camelford flood in June 1950, the

Lynmouth disaster in August 1952, further Camelford floods in June 1957 and June 1958, Porlock in July 1959, Wadebridge again in June 1963, and large parts of Somerset and east Devon in July 1968.

The main control over the quantity of rain that falls in these events is the amount of moisture supported in the troposphere – the lowest layer of the atmosphere where all our weather happens. The depressions responsible for these prolonged downpours are typically slow moving in the Southwest Approaches, so the air mass lingers for some days over the Atlantic waters just to the south-west of the UK. One measure of the humidity of the air is the dew-point, and this will be within a degree or two of the sea-surface temperature where the air has been stagnating. Thus in the middle of the twentieth century the dew-point of such an air mass would have been around 17–18°C, and with 100 per cent relative humidity would have supported a vapour pressure of 20 millibars. By the turn of the century, with dew-points about a degree higher, the potential vapour pressure in the same air-mass had reached 21.3 millibars. With a further rise in dew-point to 19–20°C by 2035 the vapour pressure may be something like 22.7 millibars, and by 2100 a typical dew-point of 21–22°C would support a vapour pressure of 25.6 millibars. Thus, other things being equal, a major West Country cloudburst which deposited 200mm of rain in the 1950s might be expected to drop 225–230mm in 2035, and 250–260mm by 2100. That is over 25 per cent more than in the 1950s.

The sharp decline in the frequency of these events since about 1970 may indicate that other factors are at work here, but we may reasonably conclude that, whether future such rainstorms occur less frequently or not, they are very likely to be more intense when they do actually happen.

The decline in the areas covered by ice and snow in the Arctic region, and over Europe in winter, is likely to speed the disappearance of severely cold periods in the UK between November and March. With appreciably less ice in the Iceland–Greenland region and in the northern Norwegian Sea in the last two decades compared with the 1950s and 1960s, northerly and north-westerly winds have already become markedly less cold than they were 50 years ago, and associated snowfall is less frequent and less long-lasting. With further losses of Arctic ice during coming decades, this process can be expected to continue. It is useful to note here that the melting of sea ice during the Arctic summer provides us with an indication of just how speculative these ideas are: several academic reports published in the late 1990s and early 2000s predicted that the global warming trend would result in sea-ice in the Arctic completely disappearing in late summer towards the end of the twenty-first century, but following the dramatic losses in 2005 and especially 2007 the date for an ice-free Arctic has now been revised to 2025. It may, of course, be revised again. European snow and ice have a positive feedback effect in respect of cold weather: snow reflects rather than absorbs heat energy coming from the sun, whereas snow-free continental surfaces absorb rather than reflect. Thus once a snow-cover is established over a large part of the continent early in the winter it puts a feedback loop in place which will tend to encourage the development and persistence of cold weather throughout the season. The absence of a European snow-cover in most recent winters has already resulted in a complete absence of severely cold days associated with easterly winds over the UK during the last 15 years. At a typical location in the English Midlands, an afternoon maximum temperature below −3°C (with an easterly airflow) was recorded on at least one day in each of 1978, 1979, 1981, 1982, 1985, 1986, 1987 and 1991, while the only instance since then, in December 1995, came in stagnant air after a particularly cold northerly outbreak.

The idea that an outflow of cold, fresh water resulting from an acceleration in the melting of the Greenland icecap will trigger a sudden cooling in the climate of lands bordering the north Atlantic has more to do with Hollywood catastrophism than with real weather disasters, but it is worth mentioning in passing. The theory is based on events which took place after the end of the last northern hemisphere glaciation when the climate of Europe and the North Atlantic suffered a dramatic reversal with glaciers re-advancing in Scotland and Scandinavia; this episode, usually called the 'Younger Dryas' but occasionally known in UK academia as the 'Loch Lomond Stadial' may have kicked in in as short a period as twenty years, and it lasted for several centuries. To ensure the necessary catastrophic effect, the Hollywood film based on the idea, *The Day After Tomorrow*, telescoped the 20–50-year onset into a couple of days! It is now generally accepted that the Younger Dryas reversal was the result of a particular sequence of events involving the retreat of the North American (usually known as 'Laurentide') icesheet at the end of the last glaciation. The geography of the Canadian Shield, bordered to the east, south, and west by high ground, and blocked to the north by the retreating ice, prevented the growing volume of meltwater from escaping. A massive freshwater lake developed, of which the present-day Great Lakes are but remnants. Eventually, the ice which dammed the 160-kilometre passage between Baffin Island and the mountains of northern Labrador was breached, allowing fresh water and melting ice to disgorge through Baffin Bay and into the north-western Atlantic. Geomorphological evidence shows that the sea level rose abruptly by some 40 metres at this time, while changes in the marine flora and fauna indicate that ocean-surface temperatures were briefly lower than they had been during the peak of the preceding glacial period. There is, however, no conceivable way that meltwater issuing from the Greenland icecap could be similarly dammed, and then ultimately escape in a similarly catastrophic fashion. Having said that, there is much we do not know about how the North Atlantic's ocean currents work, and how even a limited outflow of fresh water might affect them, and researchers are aware that the North Atlantic Drift, the extension across the northern Atlantic of the warm Gulf Stream, is subject to occasional dramatic fluctuations in its strength. Nevertheless, the conventional wisdom in the early years of the twenty-first century is that a catastrophic shift in western Europe's climate resulting from the outflow of fresh water from Greenland is very unlikely, but that smaller fluctuations are quite possible. The repercussions of such events on the character, intensity and frequency of individual weather disasters in the UK are so uncertain that there is little point even in speculating about them.

Changes in synoptic weather patterns

It is probably not wise to speculate too long, either, about the changes in the synoptic climatology of our part of Europe. Again, the most advanced climate models we have now are only scratching the surface of predicting these sorts of changes, although rapid advances in computer modelling can be expected during the next few decades. So results in this area may be extremely modest and unreliable at the moment, but they should improve markedly in future years.

For a time during the late 1990s and early 2000s it was common to hear television and radio weather presenters (often not knowledgeable about climatological matters) suggesting that the retreat of ice in the European sector of the Arctic Ocean would result in the mean Atlantic depression track migrating northwards, both in winter and in summer, and that the most likely consequence of this on the British climate would be much milder winters, wetter in northern and western Britain, but perhaps drier in the east, while summers would probably become hotter and drier with longer heatwaves, more frequent and more intense droughts, and less frequent but more intense thundery downpours. They would also talk about gales becoming more frequent because '. . . rising temperatures mean more heat energy is entering the system, and if you put more energy in, you get more energy out, typically in the form of deeper depressions and stronger winds'. They were able to point to the events of the 1990s as evidence that this was already happening because the Atlantic depressions, particularly from October to March, were more intense during that decade, and travelled deeper into the Arctic.

This demonstrates the danger of jumping to conclusions. Since the turn of the century there has been a marked reduction in the strength of the prevailing south-westerly winds in our part of the world during the winter half-year, and lengthy periods when they disappeared altogether, without quite returning to the relative quietness that typified so many (though not all) winters between 1939 and 1988. Twenty years is far too short a period to expect to detect any underlying trend in the characteristics of weather patterns in the Atlantic/Europe region resulting from the present global warming trend, especially when there are other influences at work which themselves result in fluctuations on varying time-scales. One of these fluctuations, the causes of which are poorly understood, produces a 60- to 70-year cycle in the strength of Atlantic south-westerlies: an increase in the vigour of the atmospheric circulation over the North Atlantic and Europe was apparent in the 1730s, the 1800s, the 1870s and the 1920s, so it was predictable before global warming was ever heard of that there would be a renewed resurgence towards the end of the twentieth century. Assuming that the cyclical pattern continues, winters may well turn out to be relatively quiet during the 2020s and 2030s.

Professor Hubert Lamb, probably the UK's most prominent climatologist of the twentieth century, was dismissive of those who made the simplistic link between rising temperatures and increased storminess. He demonstrated a clear association between more frequent gales and *lower* temperatures during the Little Ice Age, between the fourteenth and eighteenth centuries. And he noted that there was evidence that the Climatic Optimum – between 6000 and 3500BC – when western Europe's climate was at its warmest since the end of the last glacial period, was characterized by relatively quiet conditions with moderate rainfall and infrequent gales.

The frequency and intensity of summer heatwaves have increased markedly after 1988, while summer droughts have occurred more frequently since the early 1970s, but some aspects of the summer climate of the British Isles can also be put down to other influences, and it would be unwise to expect that every summer in the twenty-first century will be hot and dry. In that respect, summer 2007 provided an appropriately cautionary tale. As ever, a proper historical perspective is essential when considering where our changing climate is taking us, and it is foolish to try to explain each and every short-lived perturbation in terms of the underlying warming process.

Summary

We may usefully contrast what is likely to happen to weather extremes during the next hundred years with the widespread myths, often repeated in the mass media, that most of these extremes will become more frequent and more intense (see Table 10.3).

Type of extreme	What the mass-media say		What the models say		Changes so far	
	More frequent	*More intense*	*More frequent*	*More intense*	*More frequent*	*More intense*
Gales	Yes	Yes	Perhaps	Perhaps	No	No
Winter river flooding	Yes	Yes	Probably	Probably	Equivocal	No
Storm surge/ coastal flood	Yes	Yes	Probably	Probably	No	No
Flash floods	Yes	Yes	No	Probably	No	No
Snowstorms	No*	No*	No*	No	No*	No
Icestorms	No*	No*	No*	No	No*	No
Severe winters	No*	No*	No*	No*	No*	No*
Summer heatwaves	Yes	Yes	Yes	Yes	Yes	Yes
Droughts	Yes	Yes	Perhaps	Perhaps	No	No
Fog/smog	Probably	Probably	No	No	No	No
Tornadoes	Yes	Yes	Perhaps	Perhaps	No	No
Hailstorms	Yes	Yes	Perhaps	Perhaps	No	No

*Will become or are becoming less frequent

Part 2

A chronology of disaster: severe weather events in the UK from 1901 to 2008

Our knowledge of extreme weather events in the UK is very comprehensive between 1860 and 1993, thanks to the abundance of journals and other periodicals produced by both private individuals and organizations as well as by the national meteorological service. Since 1993 many periodicals have ceased publishing, detailed records and observations are less easy to acquire, and fewer individuals are inclined to spend the time and effort to examine and analyse these events for posterity, so the quality of information in the public domain has, with some honourable exceptions, declined in the past fifteen years. Before 1860 there is a good deal of documentary evidence detailing weather disasters, but most of it is not seen through the objective eyes of a meteorologist, so the descriptions tend to be qualitative rather than quantitative. As one would expect, our knowledge of these events becomes progressively less comprehensive as we travel backwards from 1860.

No claim is made here to have included every severe weather event, even during the years of plenty. But serious omissions should be few. The entries are largely confined to those events which had serious repercussions on human communities or on the natural landscape: it is therefore not a chronicle of all weather extremes. For lowland Britain, an attempt has been made to include all those occasions of gusts exceeding 90mph, 24-hour rainfall totals above 100mm, and snowfalls of 25cm or more. In upland Britain, such thresholds are passed much more frequently, and at the same time far fewer people are affected by them, so in these areas the requirements for inclusion in the list are more demanding. Even so, most gales with gusts reaching 100mph at more than one site, 48-hour rainfall totals in excess of 200mm, and 40cm snowfalls have been logged. The records for unpopulated sites, including mountain summits such as Cairn Gorm and Great Dun Fell, remote lighthouses such as the Needles and Bell Rock, and unpopulated islands such as North Rhona, have not been included.

No apology is made for the mixture of metric and imperial units in this archive. Most measurements are indeed quoted in metric units, but speeds, whether of winds or motor cars, will be most widely understood in the UK in mph. Weather observers still use knots, and few people – even meteorologists – are entirely comfortable with either metres per second or kilometres per hour as a unit of wind speed.

Similarly, county names are used for the sole purpose of helping to locate unfamiliar places, not for massaging civic egos. Thus, the traditional (or historic) counties have been used, especially in Scotland and Wales, while in a few instances the counties have been sidestepped in favour of other geographical identifiers such as 'the Lake District' or 'Snowdonia', or simply 'near Sheffield'.

The sources of the data are many and varied, including periodicals and journals, both official and private-sector, private records, newspaper cuttings, unpublished communications, and my own notebooks. Wherever possible, and that means with very few exceptions, quoted measurements have been checked against published records. No uncorroborated newspaper reports, for example, should have found their way into the chronicle. The journals and periodicals used include: *British Rainfall* and its successor publications, *Symons's Meteorological Magazine* and its successor *The Meteorological Magazine*, the *Quarterly Journal of the Royal Meteorological Society*, the *Monthly Weather Report* of the Meteorological Office, *Weather* and *Weather Log*, the *International Journal of Climatology*, and the monthly *Bulletin of the Climatological Observers' Link*.

1901–10

1901

6–9 January *Snow; severe cold*
Intense frost in southern England followed moderate to heavy snow on the 6th–8th. The cold was centred on Dorset, Hampshire, Sussex and Surrey, with –19°C recorded at Swarraton in Hampshire on the morning of the 9th. Three deaths from exposure noted; there were probably many more.

4 and 11–12 February *Heavy snow*
Snow fell heavily across much of England and Wales on the 4th, accompanied by a strong north-east wind, dislocating transport for several days. In Liverpool, snapped telephone wires came into contact with trolleybus power lines during the evening, electrocuting several people; two died and 15 were injured. The snow was deepest in upland parts of the Thames basin, with 22cm at Highmoor in the Oxfordshire Chilterns. A further snowfall on the 11th–12th resulted in a 25cm-deep cover at Norwich. Although there were periodic thaws, parts of the Midlands and East Anglia had a snow cover on 20–25 days during the month.

21–29 March *Heavy snow*
Severe snowstorm over the West Country moors on the 21st with 38cm reported at Princetown on Dartmoor. One death from exposure noted. Further snow at times in many parts of the UK on subsequent days with heavy falls locally, notably in Northern Ireland on the 28th–29th. The temperature fell to –17.2°C on the 29th at Braemar in Aberdeenshire and in Dundee.

25 July *Urban flooding; thunderstorm*
A fortnight earlier, on the 12th, 92mm of rain fell in a thunderstorm at Maidenhead, one of the greatest falls ever recorded in one hour in the UK. Thunderstorms again broke out widely between the 24th and 27th, and the one which struck London during the afternoon and evening of the 25th was particularly bad, and accompanied by hailstones 2cm across. At Camden Square 72mm fell, most of it in 150

minutes, and in the King's Cross district 300 people sought temporary accommodation after their homes were flooded; the underground railway was put out of action for 48 hours by deep floodwater.

12–13 November   *River flooding; severe gale*
A widespread and severe gale led to numerous shipwrecks in British home waters. Prolonged heavy rain resulted in floods in many parts of Wales, Yorkshire, Lancashire, Cumbria and Northern Ireland. At Todmorden it was the worst since the historic flood of 1866. Several other towns were flooded and railway lines were washed out. Calderdale was very badly hit. Rainfall totals for the two days exceeded 100mm at 27 locations, with 146mm in the Duddon Valley, Cumbria.

12–14 December   *Snowstorm*
Heavy snow accompanied by a north-easterly gale affected much of central and northern England, and also southern Scotland, especially over 150m above sea level. All railway traffic between London and northern England ceased, thousands of telegraph wires were brought down and telegraphic traffic between north and south was also suspended and there were heavy losses of livestock, mainly sheep, in upland districts. Several deaths were reported. Level snow was 40–45cm deep in Shropshire, drifts to 4m were noted in Wales, north-east England and south-east Scotland, and it was said to be the worst snowstorm in the Cheviots and Lammermuirs for over half a century. A rapid thaw between the 15th and 18th led to extensive flooding.

January to November   *Drought*
Eleven consecutive dry months allied to a long, hot summer resulted in a significant drought, with reduced crop yields, burnt pastures, cessation of traffic on some canals and dried wells. Annual rainfall deficit amounted to 30–35 per cent in East Anglia.

1902

8–9 February   *Widespread snow*
The first three weeks of February were very cold and wintry with frequent snow. The heaviest falls came on the 8th and 9th, leaving

30–35cm on the ground in parts of Lancashire, Yorkshire, Northern Ireland, the Southern Uplands and the North-west Highlands. Traffic in both Liverpool and Manchester came to a complete halt.

14 May *Late frost*
Widespread late frost with much damage to potatoes, asparagus and soft fruit, also to apple, pear and cherry blossom. Air temperature as low as –4°C in midland and southern counties of England.

10 September *Hailstorm*
A heavy thunder and hailstorm created havoc in the Maidstone and Tonbridge districts of Kent, coming mid-afternoon in the midst of the hop-picking season. Hailstones were up to 6cm across, destroying many glasshouses. Hop fields and orchards were devastated. One man was killed and several others injured by lightning. At Weybridge in neighbouring Surrey 96mm of rain fell in 90 minutes.

1903

7–8 February *Flood*
Two days of heavy rain over south-west Scotland and Cumbria resulted in severe flooding in Glasgow as the River Clyde overtopped its banks. Thousands of acres of land were submerged, in places to a depth of more than 3m, all road and rail traffic was suspended, and dozens of factories and mills were inundated. Some 6,000 workers were thrown out of work. It was the worst flood in the city for over 100 years. Several rain-gauges in the hills to the south and south-west of Glasgow recorded 100–125mm.

26–27 February *Severe gale*
A severe south-westerly gale swept much of the British Isles overnight, causing most damage in Ireland, Wales, northern England and southern Scotland. In Ireland it was considered the worst windstorm since 1839, and in northern England since either 1894 or 1886. There was widespread and serious structural damage to houses and public buildings, thousands of trees were uprooted, telegraph communication between south and north was suspended and transport severely disrupted. A train crossing the Leven viaduct,

near Ulverston, Lancashire, was overturned. The death toll nation-wide exceeded 15, and may have been as great as 30. Peak gusts included 92 mph at Southport (Lancashire), 88mph at Pendennis Castle (Cornwall) and 87mph at Holyhead (Anglesey).

28–31 May   *Thunderstorms*
Widespread thunderstorms, many severe, affected England and Wales. Several deaths from lightning strikes were reported. At Beddington, Surrey, 93mm of rain and hail fell, of which 89mm came down in just 55 mins.

8–20 June   *Severe flood*
This rainfall event appears to be unique in the UK's climatological archives, with long periods of continuous rain throughout thirteen days over southern England, but without the torrential thundery downpours normally associated with summer rainfall. The wettest area was almost exactly coincident with the Thames basin; upwards of 125mm of rain fell over 14,200 sq. km, and 150mm+ fell over 5,800 sq. km. At Carshalton, Surrey, the total for the event was 226mm. Between the 13th and 15th it rained without a single break for around 60 hours, with 90–110mm falling during this period. Floods were extensive along the Thames and Lea, haymaking was completely ruined and grain crops were badly depleted.

23 and 25 July   *Flooding*
Prolonged heavy rain fell for 9–12 hours across south-east England and East Anglia on the 23rd, with over 100mm in north-east Kent and south-west Essex; Dartford recorded 112mm. The night of 25th–26th was very wet in central London, and floodwaters entered several newspaper printing halls (off Fleet Street) resulting in the non-appearance of several Sunday papers.

10–11 September   *Severe gale*
Much damage was wrought by a severe gale, south-westerly at first, veering north-westerly, across southern England and south Wales. A feature was the extensive scorching of vegetation caused by the salt-laden wind sweeping in from the Atlantic. A major storm surge in the Bristol Channel damaged sea defences.

October   *Flooding*
This is the wettest calendar month in three centuries of rainfall records, with rainfall averaged geographically across England and Wales of 218mm. Rain fell daily in many parts of the country, and, following the wet summer, flooding was widespread and prolonged. In many regions the grain harvest was not completed, and root vegetables were left to rot. Over 100 sq. km of land lay under water at the month's end.

1904

30 January to 14 February   *Flooding*
Extensive flooding along the Thames and its tributaries, and on several rivers in Sussex and Kent, resulting from frequent bouts of heavy rain; Oxford, Reading, Maidenhead and Windsor were badly affected, with river levels at their highest since November 1894.

23–30 July   *Thunderstorms and local flooding*
Worst hit was Huddersfield where 123mm fell in five hours on the 24th.

20–23 November   *Heavy snow*
Outbreaks of snow, heaviest in Cumbria, Galloway, Dumfriesshire and in counties Tyrone and Londonderry; level depths of 25–35cm were widely recorded, with 45cm (drifts to 3m) around Keswick and Duddon Valley. Some 15–20cm on Bodmin Moor. Transport severely disrupted.

9–28 December   *Widespread fog; London smog*
Thick fog affected different parts of England and Wales for several days at a time during this period. London was continuously smog-bound from the 19th to the 23rd inclusive, disrupting road, railway and river traffic. Over 400 ships, unable to reach the Port of London, were forced to anchor in the Thames Estuary. The *Daily Telegraph* reported that the Chancellor of the Exchequer, unable to persuade any cabman to attempt the journey, was forced to walk from Downing Street to Euston Station. This episode was dubbed the 'Black Christmas of 1904'.

29–30 December   *Severe gale; storm surge*
A severe gale, initially south-westerly, later veered north-westerly, triggering a major storm surge along the east coast of England and into the Thames estuary where it was reported to be the highest since routine tidal measurements began. There was extensive coastal flooding and thousands of sheep were lost. A good deal of gale damage was reported on land.

1905

16 January   *Glaze*
Short-lived but severe episode of freezing rain in London. Following an afternoon maximum of –5°C, a fall of ice-pellets gave way to heavy rain. The glaze lasted for about three hours before temperatures rose above zero. The city's transport networks were thrown into chaos, and hundreds of people were injured after slipping or falling.

14–15 March   *Severe gale*
A severe southerly gale, later veering westerly, affected England and Wales, with strongest winds over Cornwall and Devon. Widespread wind and hail damage.

9 July   *Thunderstorms*
Severe thunderstorms accompanied by hail and rainfall totals up to 75mm hit the London area. Several people were struck by lightning and killed, including three on Hampstead Heath.

15–16 August   *Thunderstorms; flooding*
Flooding followed prolonged thunderstorms in Devon, Somerset and the south-west Midlands. At Starcross, Devon, 101mm fell in twelve hours.

1906

5–6 January   *Severe gale*
A south-westerly gale swept central and southern parts of England and Wales. Much damage to buildings and trees; several deaths reported.

11–12 March   *Storm surge*
Strong northerly gale in the North Sea almost coincided with a high spring tide. There was coastal flooding in the southern shore of the Moray Firth and along the east coast of England from Lincolnshire to Kent.

28–29 June   *Flooding*
A prolonged overnight downpour lasting approximately twelve hours deposited 50–80mm over a large area of southern England roughly bounded by London, Oxford and Cambridge. Widespread if short-lived surface flooding, including parts of the London Underground.

2 August   *Hail/thunderstorms*
Thunderstorms, accompanied by violent squalls and heavy hail, affected large parts of central and southern England and also Wales. Much hail and wind damage, notably in west Surrey and adjacent parts of Hampshire and Sussex, and also in north Bedfordshire, Huntingdonshire and east Cambridgeshire. A tornado at Guildford.

30 August to 3 September   *Heatwave*
After a mostly fine and warm summer, a major heatwave brought temperatures in excess of 30°C on five consecutive days, peaking at 35.6°C at Bawtry, near Doncaster, on the 2nd. Humidity levels were moderate but there were several deaths reported from heat exhaustion.

25–30 December   *Snowstorms*
Two major snowfalls occurred. The first, accompanied by moderate winds, affected west and south-west Scotland, Northern Ireland and most of England and Wales (except for the coastal fringe of north-east England, west Wales, Cornwall, Devon and Dorset) on the 25th–26th. A broad zone from Lancashire to the East Anglia coast collected over 20cm, and as much as 30–35cm fell in parts of Norfolk and Suffolk. The second, accompanied by thunderstorms and a severe gale, swept eastern Scotland and north-east England on the 27th–28th; these days also brought significant snow to Northern Ireland, west Wales and south-west England. In this storm, 35–45cm fell in upland parts of Durham, Northumberland and the Scottish

Borders, and also over a large area of Northern Ireland, and 60–70cm in parts of Kincardineshire and Aberdeenshire where villages were cut off for over a week. The city of Aberdeen was unreachable by road, train or telegraph for four days. A rail crash at Elliott Junction, near Arbroath, killed 22 people, and the total death toll for the week is estimated at 40–50.

1907

21–22 July   *Thunderstorms; hailstorms*
Widespread severe thunderstorms over England and Wales, often accompanied by heavy hail. Much landscape damage in Herefordshire, Breconshire, Monmouthshire and Glamorgan. Several reports of livestock killed by lightning.

15–16 October   *Flooding*
Two separate downpours, in central Scotland on the 15th (80mm in Edinburgh), and stretching from Cornwall to Yorkshire on the 16th (123mm at Kingsbridge, Devon) both resulted in serious but short-lived floods.

1908

23–26 April   *Snowstorm; floods*
A wintry spell began on the 19th, Easter Sunday, but the four days 23rd–26th produced long periods of heavy snow across south-east England, Wessex, the Thames Valley, the Midlands and East Anglia. The most dramatic day was the 25th which produced the deepest snow of the entire twentieth century across much of Hampshire and the Isle of Wight, Berkshire and south Oxfordshire. Level snow lay to a depth of 32cm on Alderney in the Channel Islands, 35cm at Freshwater on the Isle of Wight, 38cm at Southampton, 40cm at Oxford, 70cm at Abingdon, and it is estimated that between 70 and 90cm fell in and around Kingsclere (Hampshire). A rapid thaw from the 27th onwards led to severe floods along the Thames, Kennet, Cherwell, Great Ouse and their smaller tributaries. The flood at Buckingham was said to be the worst for half a century. Loss of life in the snow and the floods estimated between ten and twenty.

3 June   *Flash flood*
Localized downpour above Skipton, with 103mm in 150 minutes at Upper Barden Reservoir, although much more probably fell locally. Considerable landscape damage in Airedale, Wharfedale and Wensleydale; roads and bridges washed away.

18–21 October   *Flooding*
At Portland, Dorset, 238mm fell during this four-day period, of which at least 165mm fell in five hours on the morning of the 21st. Weymouth also had over 100mm, and on the 19th Treharris Reservoir in Glamorgan collected 122mm. There was serious flooding in both the Welsh valleys and in the Weymouth–Portland district.

27–30 December   *Snowstorms*
Without approaching the severity of the post-Christmas snows of 1906, this spell was nonetheless severe and disruptive with over 30cm of snow, often heavily drifted by the strong east wind, in places as far apart as Dorset, mid-Wales and Aberdeenshire. At Forfar, Angus, an average depth of 70cm was reported, and many villages in eastern Scotland were cut off until 5 January.

1909

14–19 January   *Snowstorm; floods*
Heavy snowfall with much drifting, chiefly on 15th–16th, in western Scotland, northern England and Northern Ireland, resulting in blocked roads and railways. Rapid thaw from the 17th caused widespread floods, said to be the worst in Northern Ireland and southern Scotland for over 25 years.

26 February–7 March   *Snowstorms*
Very severe spell, the temperature as low as −18°C at Marlborough, Wiltshire, on the 3rd. Several heavy snowfalls in various parts of the UK, but worst hit were Kent and East Sussex on the 3rd with 25cm generally, and as much as 55cm at Tonbridge.

27–28 September   *Flooding*
Approaching 90mm fell in west Glamorganshire, with 50–75mm

over a wide area. Roads and bridges washed out, many homes flooded, and 200 people were housed in temporary accommodation. Several deaths reported.

26–28 October *Flooding*
Southern and south-eastern counties of England were affected by repeated bouts of heavy rain, and in east Kent over 150mm fell, with 172mm in the Folkestone district of which 91mm fell on the 28th alone.

1910

21–30 January *Snowstorms*
Heavy snow with severe drifting at times affected chiefly eastern Scotland and north-east England with level depths widely over 30cm (53cm at Wearhead in the northern Pennines). Several lives lost, and heavy livestock losses. Balmoral recorded –23°C on the 28th.

5–10 June *Thunderstorms, floods*
A period of widespread thunderstorms over England and Wales. On the 9th these storms were accompanied by rain of unprecedented violence in the Oxford and Reading districts, leading to severe short-term flooding in the two urban areas. Unofficial gauges collected about 140mm at both Wheatley and Kidmore End, and although these are regarded as over-estimates, it is likely some 100–120mm actually fell. Many glasshouses destroyed by hail.

13–19 December *Storm surge*
Several days of heavy rain and south-westerly gales culminated in a substantial storm surge in and around English Channel and Bristol Channel coasts; much damage to sea defences, flooding on the Somerset Levels, Chesil Bank breached, Pagham Harbour (reclaimed in the mid-1870s) inundated (subsequently left unreclaimed), and Selsey cut off; coastal flooding at Neath. Also a good deal of flooding on the Thames, Severn, Great Ouse and Trent. Remains of Crickhowell castle demolished.

1911–20

1911

25–31 May   *Thunderstorms*
This thundery period culminated in a ferocious storm in the late afternoon of the 31st, and is known as the 'Derby Day Storm' because it broke as proceedings at Epsom racecourse were drawing to a close. In the London area alone seventeen people were killed, and flooding was severe, if of short duration, in the metropolitan area. At Banstead, Surrey, 92mm of rain fell in three hours, of which some 80mm fell in 45 minutes.

23–24 June   *Flooding*
A 36-hour long downpour in eastern England and eastern Scotland deposited upwards of 50mm over a large area, and 110mm in Upper Weardale (Durham) and at Howich Hall, near Alnwick (Northumberland). There was a good deal of flooding in many parts of eastern Britain, but Northumberland and Durham were particularly badly hit. There was also flooding in north Wales, notably on the River Ogwen, and a huge landslip blocked the Nant Ffrancon Pass.

July and August   *Heatwave*
Repeated heatwaves during this two-month period, sometimes accompanied by high humidity, meant that July and August together comprised the hottest high summer in the Central England Temperature (CET) record between 1659 and 1983. The CET of each month was 18.2°C, a figure which has only been exceeded in both July and August in 1995. Although public health had improved dramatically since the last great hot summer of 1868, there was still appreciable loss of life as a result of gastric and other disorders. The heat and humidity probably also contributed to repeated episodes of civil and industrial unrest.

4–5 November   *Severe gale; storm surge*
A severe south to south-westerly gale swept most of the UK, but was particularly bad in north-west England, southern and central Scotland. Highest gusts were 80mph at Southport (Lancashire) and

90mph at Eskdalemuir (Dumfriesshire). Woodland was devastated in Argyllshire, Dunbartonshire and Perthshire, and general gale damage was reported in all parts of the country. A storm surge caused coastal flooding on the Firth of Clyde.

January to October *Drought*
An extended drought, exacerbated by the heat and sunshine of the summer, affected much of England, eastern Scotland and parts of Northern Ireland. Averaged over England and Wales, rainfall was 30 per cent below normal during this period, and in the English Midlands, and also in Fife and Angus, the deficit amounted to just over 40 per cent. Burnt-up pastures, low grain yields and a poor fruit harvest were the results.

1912

17–18 January *Heavy snow; freezing rain*
Heavy freezing rain in Kent, Surrey and London, but heavy snow in other parts of England. Many reports of 25–35cm across the Midlands and Lancashire; 53cm at Hoar Cross, near Uttoxeter (Staffordshire). Little drifting. Much damage to trees.

4 March *Squall line*
Violent squall noted in many parts of southern England, lasting less than twenty minutes but causing extensive damage to property and to trees. Gust of 98mph at Pendennis Castle, Cornwall. At Swansea (Bwlch Waterworks) 92mm of rain fell.

25–26 August *Flooding*
The great Norfolk flood: 205.5mm fell at Brundall, near Norwich, in little more than 24 hours. This was one of the great floods of the twentieth century (see Chapter 6).

June, July and August *Poor summer*
August was the worst month of what was arguably the worst summer of the twentieth century. It was the coldest August in the Central England Temperature record which began in 1659, the wettest August in the England and Wales rainfall record (began

1727), and the dullest August in the England and Wales sunshine record (began 1881). The effect on agriculture was extremely serious, and only a relaxation of the rains in September and October prevented it from being a complete disaster.

December *Flooding; gales*
Widespread river flooding followed repeated bouts of heavy rain. Seathwaite (Cumbria) recorded 145mm on the 13th alone. Worst hit were the Conway valley in north Wales, the Tay at and above Perth, and later the Trent, Nene, Great Ouse, Thames and Severn. A fierce gale overnight 25th–26th also caused much damage in southern Britain, also triggering a storm surge along the English Channel coast. Pendennis Castle recorded a gust of 98mph.

1913

10–12 January *Snowstorm; gales*
Heavy snow, badly drifted, affected northern England, southern and central Scotland, and seriously disrupted road and rail traffic. Depths of 25–35cm were widely observed; in Perthshire 44cm was reported at Crieff and 78cm at Kenmore.

17 September *Flood*
The Doncaster storm: an extremely localized downpour deposited 155mm over the town in fourteen hours. Practically every house in Doncaster – certainly every one with a basement – was flooded.

27 October *Tornadoes*
A tornado travelled northwards across Glamorgan, from Barry on the coast to Taff Vale, where six people in the mining village of Edwardsville were killed, thence to the Cynon Valley. This was the UK's most deadly tornado. Over 200 homes were so badly damaged that they had to be demolished. A tornado, possibly the same one, and almost certainly associated with the same thunderstorm system, cut a swath through woodland at Wistanstow, south of Church Stretton (Shropshire), and caused damage in the Wem and Whitchurch districts, and also in Cheshire to the south of Runcorn.

1914

7–8 May *Flooding*
A steady downpour lasting between 24 and 30 hours hit the northern half of Scotland, with as much as 105mm at Drumnadrochit and 110mm at Ferness, south-east of Nairn.

23–27 May *Late frosts*
Widespread air frost on five consecutive nights damaged vegetable and fruit crops. Braemar (Aberdeenshire) logged –7.2°C on the 25th; even in southern England –3°C occurred widely.

14–18 June *Thunderstorms; flooding*
Severe thunderstorms hit southern England on the 14th with torrential rain, notably in a zone stretching from east Berkshire to south London; 94mm fell at Richmond Park in less than three hours, and 88mm at Staines in five hours. Seven deaths noted from lightning strikes, and there was much local flooding. On the 18th an extraordinary (unmeasured) deluge caused floods and landslips in the Ruffside district of north-west Durham; there were several lightning fatalities.

1–3 July *Thunderstorms; flooding*
Three days of very thundery weather, with violent storms particularly on the 1st when several deaths were reported. Flooding was severe, but localized and short lived. Tibberton Court in Gloucestershire recorded 99mm, mostly in two hours. Dramatic flood in Bristol.

8 August *Flooding*
Serious flooding in upland districts of England and Wales, especially in Cumbria and Snowdonia; 120mm fell at Seathwaite in Borrowdale, 117mm at Blaenau Ffestiniog and 161mm at Pen-y-Gwryd in the shadow of Snowdon.

28 December *Severe gale*
Damaging gale in south-east England and along the Channel coast; gust to 83mph at Dover.

29 November to 5 March 1915   *Flooding*
The winter quarter (December, January, February) was the wettest in the entire England and Wales rainfall series which began in 1727. Averaged over England and Wales, 423mm of rain fell, approaching twice the normal amount. Repeated bouts of heavy rain led to chronic flooding in most parts of the UK, and the floodplains of England's major rivers were under water for most of the season. A new spring bubbled through the floor of Salisbury Cathedral, flooding most of the building on 5–6 January to a depth of several centimetres.

1915

22 January   *Heavy snow*
Heavy snow to a depth of 20–30cm affected much of Kent, Sussex, Surrey, London and south Essex. Rail and road transport largely suspended for 24 hours, and telephone and telegraph services were also badly hit.

6 May   *Thunderstorms; flooding*
Severe localized thunderstorms over inner London, the centre of heaviest rainfall lying between the City and King's Cross. At Finsbury 79mm fell in 90 minutes. Hundreds of properties were flooded, and the Underground system was suspended for several hours.

4 July   *Thunderstorms; heavy hail*
Fierce thunderstorms broke out widely across east Wales, the Midlands, Lincolnshire and Yorkshire, and many were accompanied by damaging hail. At Abergavenny 56mm of rain fell in 30 minutes, and near Woodstock 74mm fell in 65 minutes. Hailstones 4 to 5cm across were reported at Shipham (Somerset). Bristol was badly hit, as it had been a year before.

16 July   *Flooding*
A prolonged downpour deposited 50–100mm over much of Staffordshire and Derbyshire, and adjacent parts of Warwickshire and Leicestershire. Heaviest falls occurred in the area bounded by Burton-upon-Trent, Loughborough and Derby, and these (plus

Nottingham) were the towns where flooding was most extensive; 101mm fell at Loughborough.

25–26 September   *Flooding*
The southern shore of the Moray Firth, especially between Inverness and Buckie, has a history of destructive, if mercifully rare, floods. On this occasion the deluge lasted 40 hours, and over 100mm of rain fell over practically the entirety of the historic counties of Nairn and Moray, together with adjacent parts of Banffshire and Inverness-shire, and the heaviest fall of all occurred at Dalcross Castle, a few miles east of Inverness, where 179mm was collected. The downpour was accompanied by a northerly gale. The Rivers Spey and Findhorn took most of the water, with devastating floods in the lower portions of their catchments. On the Highland Railway long stretches of track were washed away and sixteen bridges and culverts destroyed. Over 20 sq. km of farmland lay under water for almost a week, and recently reclaimed land at Buckie was lost. This was widely described as the worst inundation in the district since the historic flood of 1829.

27–29 October   *Flooding*
Just round the corner from the Moray Firth, another prolonged deluge caused extensive flooding in Aberdeenshire, Kincardineshire, Angus and Perthshire. Over a 48-hour period, 132mm fell at Durris and 131mm at Crathes, both near Banchory.

9 November   *Flooding*
The Spey and Findhorn catchments were flooded for a second time in two months, setting back recovery work from the earlier disaster by several weeks. At Alvie, near Aviemore, 97mm fell, and at Knockando distillery (Morayshire) 94mm.

27 December to 1 January 1916   *Severe gale*
Widespread gales swept the country during the last four days of the year, with wind damage reported from all corners of the country. Peak gusts included 92mph at Pendennis Castle, near Falmouth, and on St Mary's (Isles of Scilly), and 90mph at Plymouth, all on the 27th, and 87mph at Southport (Lancashire) on the 1st.

1916

16 February   *Severe gale*
Widespread gales, most damaging over north Wales, northern England, southern Scotland and Northern Ireland. Gusts to 90mph at Holyhead and 83mph at Southport.

Late February, March   *Heavy snow*
This was an exceptionally snowy period in upland parts of Britain, including Wales and the West Country moors; level snow lay 50–150cm deep, with some reports of drifts 15m high. Moorland farms and villages were isolated for weeks, there were considerable losses of livestock and upland railway lines closed down until early April.

27–28 March   *Snowstorm; severe gale*
An intense secondary depression tracked from Cornwall to Suffolk, and in its immediate wake heavy snow fell, while the northerly wind increased to storm force. Thousands of trees were uprooted (95 in Kensington Gardens in London alone), telegraph poles broken, roads and railways blocked, telephone, telegraph services disrupted, tramlines brought down, and there was also a good deal of structural damage. Worst hit were the south-east Midlands and northern Home Counties, especially Northamptonshire, Buckinghamshire, Bedfordshire and Hertfordshire. Snow depths in these areas averaged 20–30cm, but there was massive drifting. Peak gusts included 87mph at Dover and 74mph at Kew Observatory – this was a record for the site until 1973. Serious flooding followed a rapid thaw, especially on the Nene and Great Ouse, as March gave way to April.

7–8 July   *Flooding*
A prolonged downpour lasting 48–60 hours hit much of eastern and central Scotland; wettest areas were the Lothians, Fife, Stirlingshire, Perthshire and Angus, with around or rather more than 100mm. At Perth the fall amounted to 140mm and at Dundee 118mm.

29 August   *Flooding*
A 12–18-hour downpour affected much of southern, central and

eastern England; worst hit areas were south Devon, Dorset, Wiltshire and also Suffolk. Torquay collected 108mm of rain, and Donhead St Mary (near Shaftesbury) 111mm.

11 October   *Flooding*
A stream of warm and humid air spent itself on the western highlands of Scotland where Kinlochquoich recorded 208mm of rain in 24 hours. The West Highland Railway and local roads were washed out in many places and were impassable for many days, leaving Fort William completely cut off. Glen Nevis became a huge lake for several days, and many hundreds of sheep were lost. There was further heavy rain on the 13th and 14th.

25 October to 7 November   *Severe gales*
This was a very stormy period with frequent gales. Peak gust of 93mph at Pendennis Castle was recorded on the 28th. The steamer *Connemara* foundered off north-west Ireland.

1917

January to April 1917   *Long cold and snowy winter*
The period mid-January to mid-February had a Central England Temperature close to zero, and much of the country was snowbound for three to four weeks, and snow lay 30cm or more deep in north-east England and eastern Scotland during this period, with 45cm at Bellingham, Northumberland. The winter quarter was the coldest during the period 1896–1939 inclusive. The lowest individual temperature was –20°C overnight 5/6 February at Benson in Oxfordshire. It was an exceptionally difficult winter on the land, made worse by the continuation of wintry weather throughout March and the first half of April. Unprecedented snowstorms hit Ireland during the first few days of April, and there were heavy, though more localized snowfalls in many parts of Britain too. A minimum temperature of –15°C was recorded at Newton Rigg, near Penrith, on the morning of 2 April.

16 June   *Thunderstorm, hail, flooding*
A severe thunderstorm broke over the western and northern out-

skirts of London during the afternoon, with over 50mm of rain in a zone extending from Richmond to Finsbury Park. Cam House in Kensington collected 118mm in just over two hours. Large hailstones caused much damage, and flooding was severe if localized.

28–29 June   *Flooding*
Somerset and south Wiltshire were visited by a record-breaking downpour which lasted about eighteen hours and which led to destructive floods, notably in the small towns of Gillingham and Bruton. In Bruton itself, 243mm of rain fell at Sexey's School, and 215mm at King's School.

29 July to 3 August   *Flooding*
Repeated downpours over this six-day period hit south-east England and East Anglia, and Kent was dealt a particularly hard blow with a total rainfall of 262mm at St Thomas' Hill, Canterbury, 224mm at Chilham, 220mm at Molash and 215mm at Ospringe. Flooding was very bad in Canterbury itself, and also across parts of Romney Marsh. At the time there was much speculation that the downpour had been caused by the percussive effect of heavy shelling in Flanders.

23–24 August and 28 August   *Severe gale*
Two rare summer gales caused a good deal of damage along the English Channel coast. At Dover a gust of 78mph was reported on the 28th.

24–26 October   *Severe gale*
A severe gale raked central and northern England and southern Scotland causing widespread damage. Peak gusts included 92mph at Eskdalemuir and 90mph at Southport.

23–30 November   Flooding
A phenomenal eight-day rainfall of 578mm (a year's worth of rain in London) was recorded at Kinlochquoich, above Glen Garry, in north-west Inverness-shire. Destructive floods were reported from Loch Awe in Argyllshire and at Beauly, just west of Inverness.

16–17 December   *Severe gale*
The Newquay lifeboat was wrecked in a ferocious south-westerly gale. St Mary's (Isles of Scilly) logged a gust of 99mph.

1918

1–16 January   *Severe cold; heavy snow; floods; storm surge*
Intense cold; several deaths reported from exposure; on the 8th the temperature remained below –4°C all day over much of the country, and as low as –6°C at Little Massingham (Norfolk) and Kingussie (Inverness-shire). There were heavy drifting snowfalls too, notably in northern Scotland on the 7th, resulting in the closure of the Highland Railway for a week, and across England and Wales on the 15th which was followed by a rapid thaw and widespread flooding. Heavy livestock losses. The thaw was accompanied by a south-westerly gale and storm surge along the English Channel coast; maximum gust 78mph at Plymouth.

28 February to 2 March   *Gales and storm surge*
A north to north-easterly gale with gusts to 76mph at Aberdeen and Eskdalemuir was associated with a major North Sea storm surge which caused coastal damage in East Anglia, the Thames Estuary and east Kent.

17 May   *Thunderstorms; flash floods*
Widespread thunderstorms and heavy hail; worst hit areas were Staffordshire, Lincolnshire and Bedfordshire/Buckinghamshire. Highest measured rainfall was 82mm in under three hours at Lidlington (Bedfordshire), but it is believed 125–150mm may have fallen in the nearby Eversholt/Tingrith district. Lanes described as turning into raging torrents with water up to the tops of the hedgerows. Long stretches of the estate perimeter wall at Woburn Abbey collapsed. Flash flooding also reported from Chesham (Buckinghamshire).

16–22 July   *Thunder/hailstorm*
This long thundery period started with a brief but fierce hailstorm which smashed glasshouses, stripped trees and flattened crops in

many parts of Surrey; 28mm of rain fell in eleven minutes at Purley, and some hailstones were 6cm across.

1919

1–2 January   *Severe gale*
A severe gale swept western and northern Britain, with a maximum gust of 87mph at Southport, Lancashire, on the 2nd. A naval vessel foundered off the Hebrides with the loss of twenty lives.

4–5 January   *Heavy snow; thaw flood*
Heavy wet snow affected the Midlands, Wales and northern England, especially upland areas, bringing down hundreds of kilometres of telephone and telegraph lines. 36cm fell at Buxton. Widespread flooding followed mid-month as heavy rain added to a rapid thaw.

28 January   *Heavy snow*
Deep snow affected the northern Home Counties, East Anglia and parts of the Midlands, with disruption to road and rail transport and to communications lines. 28cm fell at Hemel Hempstead (Hertfordshire).

30 March   *Heavy snow*
A heavy snowfall over London and the south-east occasioned much disruption. Level depth reached 32cm at Lewisham, south London.

27–28 April   *Heavy snow; thaw flood*
A phenomenal snowstorm for so late in the season hit much of southern, eastern and central England, but it was particularly severe in the northern Home Counties where several places in Essex, Hertfordshire and Bedfordshire had 30–40cm. A rapid thaw led to much flooding.

12 June   *Flooding; hailstorm*
A five-hour thundery downpour led to severe flooding in Ayrshire, Renfrewshire, Lanarkshire and Dunbartonshire; 91mm fell at Ayr. A violent hailstorm hit Branxholme, near Hawick, in the Scottish

Borders. There was also heavy rain and flooding in mid-Wales where 104mm of rain fell at Devil's Bridge.

26 August *Flooding*
Serious floods hit Sutherland, Caithness and Orkney following a lengthy downpour which dropped 98mm of rain on Kirkwall and 97mm on Tongue.

8–16 November *Severe frost; heavy snow*
By far the most severe spell of weather to hit the UK in November coincided with the first anniversary of the Armistice; both the cold and the snow were most extreme in Scotland. The temperature fell to −23.3°C at Braemar early on the 14th, and climbed no higher than −10°C during the 14th and −12°C on the 15th. Snow lay 43cm deep in Braemar and 20cm deep in Edinburgh.

1920

9–15 February *Flooding*
Heavy rain – orographically enhanced – fell across western and central Scotland, north-west England and north-west Wales, result-ing in considerable flooding on the rivers of these areas. On the 9th alone, 173mm of rain fell at Dungeon Ghyll in Cumbria, 160mm at Kinlochquoich, Inverness-shire, and 130mm at Grasmere, Cumbria.

16 March *Heavy snow*
A fall of 25–30cm fell over much of central and eastern England, with 30–35cm in many parts of Northamptonshire, Lincolnshire, Rutland, Leicestershire and Nottinghamshire.

25–29 May *Thunderstorms; flash flood*
Several days of very thundery weather culminated on the 29th in the Louth Disaster. At nearby Elkington Hall 117mm of rain fell inside three hours, and the deluge may have exceeded 150mm locally. A blockage in the River Lud gave way under the weight of water, and a devastating torrent swept through the small town of Louth, killing 22 people and demolishing several houses. The river rose 4.5m in less than fifteen minutes. On the same date 82mm of rain fell in two

short sharp downpours at Leyland (Lancashire) resulting in severe local flooding.

June to August   *Poor summer*
This was the coldest and one of the dullest summers of the twentieth century, resulting in a poor harvest with low yields and late ripening of crops.

17 August   *Flooding*
A prolonged downpour delivered over 50mm of rain throughout the Central Belt of Scotland, with 108mm at Bathgate, 105mm at Linlithgow, and 104mm at Harperrig in the Pentland Hills. Destructive floods hit many towns and villages, not least both Edinburgh and Glasgow, and coal mines were put out of action due to water ingress.

1–4 October   *Flooding*
Over 100mm of rain fell in this four-day period over the entire catchment of the Scottish River Dee resulting in extensive and destructive flooding on Royal Deeside and in the city of Aberdeen. Aitnach, near Ballater, recorded 184mm, and Balnaboth, near Kirriemuir, 180mm. There were heavy livestock losses and the harvest was ruined, with thousands of sheaves of corn floating out to sea. It was reckoned, locally, to be even worse than the great flood of 1829.

3 December   *Severe gale*
Many vessels lost at sea in British waters during a severe southwesterly gale. Gusts reached 84mph at Southport and at Balmakewan, Kincardineshire. A tram was blown over near Halifax, injuring its twenty passengers.

5–17 December   *Heavy snow*
This very cold spell centred on a heavy snowfall across central and southern England on the 11th, with 35cm reported at Clacton in Essex, and 30–38cm at Salcombe in south Devon.

1921–30

1921

August 1920 to December 1921 *Drought*
Below-average rainfall persisted from late summer 1920 until the end of 1921, and the calendar year 1921 was the driest ever recorded in eastern, central and southern England. This, allied to high temperature, very low humidity and abundant sunshine for much of the summer led to extreme drought stress and major domestic water shortages from July 1921 until early the following year. Lowest annual rainfall for 1921 was 253mm at Southminster in Essex (the oft-quoted 236mm at Cliftonville in Margate was the result of a faulty gauge); large parts of Essex and Kent had less than 50 per cent of their normal rainfall over the year, which is probably a unique occurrence anywhere in the UK.

1 and 5–6 November *Severe gales; storm surge*
Widespread gales. That of the 1st was a north-westerly one and triggered a major North Sea storm surge with flooding along the coasts of Lincolnshire, East Anglia and the Thames Estuary; there was also flooding in London Docks and as far upstream as the Victoria Embankment.

19–21 and 27–28 November *Persistent smog*
Widespread fog, accompanied by much smoke, enveloped Britain's conurbations between the 19th and 21st, including London, and London was also affected a week later. There were many road accidents, and rail and sea traffic was also badly affected.

16–17 December *Severe gale; storm surge*
Another north-westerly gale and North Sea storm surge; serious flooding in Hull. Heavy losses in coastal waters.

30 December *Severe gale*
A south-westerly gale caused extensive gale damage on land, and several vessels foundered around British coasts. The highest recorded gust was 78mph at Holyhead.

1922

8 March *Violent gale*
A short-lived but violent westerly gale hit south-west England; St Mary's on the Isles of Scilly recorded a mean wind speed over 80mph lasting almost an hour either side of 4am GMT, with a peak gust of 108mph. Pendennis Castle had a peak gust of 103mph and Plymouth 96mph. Damage was widespread on land and there were several losses at sea.

14–15 May *Flooding*
Prolonged heavy rain in the South-west Highlands of Scotland led to a good deal of flooding on rivers in Argyllshire and Perthshire. Over the 48-hour period, Glencammel on Mull recorded 179mm, Ardgour House 178mm, Glen Etive 145mm, Dalmally 136mm, and Loch Awe 134mm.

21–25 May *Hailstorms*
A thundery period, with damaging hailstorms in Wensleydale on the 21st and in Kent and Surrey on the 25th; on the latter occasion hundreds of glasshouses were badly damaged.

June to August *Poor summer*
Just two years after the coldest summer of the century came the second coldest. Summer 1922 was particularly noted for a lack of sunshine as well as for the persistent coolness. Agriculture suffered in all parts of the country.

6–7 August *Flooding*
An extended downpour which hit a broad zone in the middle of England from Dorset up to Yorkshire produced serious flooding in south Yorkshire, Derbyshire, Nottinghamshire, Leicestershire and Rutland. Heaviest rain occurred in Nottinghamshire where, over a 28-hour period, 146mm fell at Ollerton, 140mm at Scrooby, 137mm at Bawtry Hall and 129mm at both Worksop and Retford.

1923

18–24 February   *Heavy snow*
Frequent falls of snow affected Scotland, northern England and the north Midlands; it lay 20–30cm deep over a large part of Derbyshire and Yorkshire, disrupting road and rail traffic.

24 May   *Late frost*
Widespread frost caused serious damage to vegetable and soft fruit crops, especially in eastern and central England where the temperature fell to –2 or –3°C at several places.

7–15 July   *Violent thunderstorms; flash floods*
The thunderstorm on the night of 9th–10th in London and the south-east was widely described as the most dramatic in living memory. Eton College chapel was destroyed by fire after a lightning strike, and dozens of houses were also struck. Seaford in Sussex collected 103mm of rain, and there was general flooding in the town, and also in Newhaven and parts of the London area. A few days earlier a cloudburst caused severe damage to roads, bridges and the railway at Carrbridge in Inverness-shire; no local rainfall data are available.

12–18 November   *Flooding; gales; heavy snow*
Oughtershaw in West Yorkshire recorded 90mm of rain on the 12th, Dungeon Ghyll in Cumbria logged 153mm in 48 hours on the 12th/13th. There was extensive flooding in north-west England and north Wales; Sale Priory in Cheshire was flooded to a depth of 3.5m, Portmadoc and Bury were also badly hit, and 300 houses were inundated in Clitheroe. High winds also caused some damage, notably on the 15th when Southport recorded a gust of 82mph, and also on the 15th north-east Scotland suffered a severe snowstorm, with 76cm at Braemar.

24 November to 8 December   *Widespread fog*
This was probably the foggiest episode of the 1920s, with serious disruption to air and sea traffic. On the 25th the whole of England and Wales practically came to a standstill.

25–26 December *Heavy snow*
Heavy snowfalls affected much of Scotland and northern England; 20cm fell in Glasgow, the heaviest fall there since 1891, while depths of 60–90cm were reported from Aberdeenshire.

1924

28–29 February *Snowstorm*
North and north-east Scotland were snowbound for several days following a blizzard which deposited 25–35cm of snow – 38cm at Achnashellach in Wester Ross. The snow was blown into deep drifts, paralysing road traffic and stranding several trains.

17–31 May *Thunderstorms; floods*
This very thundery period produced two notable events. On the 20th violent thunderstorms accompanied by heavy rain hit Bedfordshire and neighbouring counties, causing much damage to glasshouses and crops; 97mm of rain fell at Shefford. On the 31st heavy rain fell widely with some exceptional totals in Worcestershire, Shropshire and Flintshire, and also around Sunderland. 135mm fell at Hanmer (Flintshire), 123mm at Ludlow (Shropshire) and 104mm at Sunderland; serious flooding affected large parts of the north-west Midlands, Cheshire, north-east Wales, and also Wearside and Tyneside.

22 July *Thunderstorms*
Severe local storms hit Surrey and London with violent hail. As much as 103mm of rain fell at Wisley Gardens in less than two hours.

18 August *Flooding*
A phenomenal localized downpour centred just west of Bridgwater, Somerset, caused a destructive flash flood with much damage to property, livestock and crops. A total of 239mm of rain fell at Brymore House, Cannington, most of it falling in under five hours.

21 September *Severe gale*
Widespread gale damage in north-west England, north Wales and Northern Ireland. A gust of 87mph was recorded at Southport.

7–8 October *Flooding; gales*
Severe flooding followed heavy rain in Cornwall. Constantine recorded 92mm of rain and Porthcurno 89mm. Gale damage in the Channel Islands was described as the worst for several decades.

22 December to 4 January 1925 *Flooding; gales*
Two weeks of very rough weather brought repeated gales and frequent rainfall, heaviest over upland districts of western Scotland, north-west England and Wales, resulting in widespread flooding. The Thames was badly hit, the peak of the flood wave between 1.5 and 2m as it travelled downstream, reaching Teddington on the 6th. Peak gusts included 90mph at Lerwick on the 25th, 80mph at Edinburgh and Rosyth on the 27th, 81mph at Farnborough on the 1st, and 82mph also at Farnborough on the 2nd. A further gale swept Scotland on 14 January with a gust of 96mph at Lerwick.

1925

16 April *Severe gale*
Unusually late in the season, this gale caused a good deal of structural damage in Lancashire and north Wales, and the R33 airship was torn from its moorings. The highest gust was 83mph at Fleetwood.

22 July *Thunderstorms; hailstorms*
Severe storms hit the London area with dozens of lightning strikes, and there was much damage to windows, roofs and glasshouses in north-east Kent and south-east Essex from hailstones 6cm in diameter. Severe squalls and some tornadic activity were also reported.

16–23 November *Smog*
Widespread fog affected many parts of the country at times during this period, with heavy pollution in industrial areas. Thick smog cloaked the Glasgow district with scarcely a break throughout the seven days ending on the 23rd.

25–26 December   *Heavy snow*
Heavy snow fell in northern Scotland, reaching a depth of 45cm at Lairg, but it soon melted.

30 December to 10 January 1926   *Severe gale; flooding*
Considerable wind damage reported from Wales and north-west England during a short-lived gale during which a gust of 88mph was recorded at Southport and one of 85mph at Holyhead. A good deal of flooding was also reported on river systems in Scotland and northern England due to a combination of heavy rain and rapid thaw. There were significant floods on the Thames as well.

1926

16–17 January   *Heavy snow*
Snow fell widely over England and Wales. A depth of 30cm was noted at Farnborough, Hampshire.

4 and 10 March   *Severe gales*
Widespread gales with gusts to 80mph occurred on both dates; Fleetwood reported 84mph.

17–18 July   *Thunderstorms; hail; flooding*
Severe storms moved northwards from the Channel, the worst affected areas being west Dorset, east Devon, Somerset and much of Wales. Pembroke Dock recorded 116mm of rain, and Abergwesyn, Ceredigion, had 154mm. Hailstones 3–5cm in diameter were noted near Chard in Somerset, near Redruth in Cornwall, at Haverford-west in Pembrokeshire and at Woolacombe in Devon. The River Wye rose 3.5m above normal at Ross. Flash flooding caused damage at Lyme Regis, Seaton, Sidmouth, Axminster, Woolacombe, Wells, Llandovery, Llanwrtyd Wells, Haverfordwest and Mold.

4 November   *Severe gale; storm surge; flooding*
Strong south to south-westerly winds, gusting to 84mph at Edinburgh, contributed to a significant storm surge along the west coast of Scotland; with rivers in spate following heavy rain there was a good deal of flooding in tidal stretches of these rivers. Glasgow,

where the tide was the highest for 44 years, was particularly badly affected. River flooding was also reported from Wales and north-west England where 117mm of rain fell at Buttermere.

1927

28 January *Destructive gale*
Sometimes known as the 'Paisley storm', thanks to the first-ever report on the Scottish mainland of a gust greater than 100mph emanating from the Coats Observatory in the Renfrewshire town, this violent gale affected a much wider area, with widespread storm damage to property in the urban areas of the Central Lowlands, and to the forests of western Scotland, notably around Oban and Loch Fyne. Renfrew aerodrome reported a gust of 102mph, Paisley 104mph, and Tiree, in the Hebrides, 108mph. The confirmed death toll was eleven, but in practice was probably between 20 and 30, and hundreds were injured.

June to September *Poor summer; frequent flooding*
The four-month period was characterized by repeated bouts of heavy rain and a marked paucity of sunshine, and it therefore comprised the most unfavourable season for the ripening and harvesting of grain and root crops since 1879. The wettest periods were between 6 and 14 July when severe thunderstorms affected many parts of Britain; on the 11th alone 102mm of rain fell in Wolverhampton, 100mm at West Bromwich, and 87mm in less than 45 minutes at Holland House, Kensington. Strinesdale reservoir, above Oldham, collected 107mm, mostly in three hours, on 14 July, while 152mm fell at Blaenau Ffestiniog on 27th August. Eastern Scotland and north-east England were badly hit on 22 September, with 97mm in the Pentland Hills and 95mm at Stonehaven.

28–29 October *Severe gale; storm surge; heavy rain; flooding*
A severe south-westerly gale, gusting to 96mph at Southport, produced an unprecedented storm surge along the Lancashire coast; flooding was particularly bad along the entire Fylde coast, and large parts of Fleetwood were under water.

21 December *Freezing rain*
A severe glazed frost affected many parts of England and Wales, and was particularly serious in London and the Home Counties where hundreds of minor street accidents occurred.

25–27 December *Snowstorm*
For intensity, geographical extent, disruption to transport, livestock losses and landscape damage, this was easily the most severe snowstorm since March 1891, and arguably the most extreme of the entire twentieth century. Snow set in across the Midlands and Wales on Christmas afternoon while rain fell in southern districts, but rain was replaced by snow even in southern counties that night. Over 30cm fell over most of the area south-east of a line from Norwich to Dorchester, and also over much of Dartmoor and Bodmin Moor, and 60–75cm fell over the North Downs and the Weald. Drifts of 6 to 8m were widely reported. A-roads were blocked for up to a week, and B-roads for two to three weeks. The snow had a high water content, and clung to telegraph poles and wires, bringing down hundreds of kilometres of lines, as well as damaging thousands of trees. Rainfall equivalent of 75–80mm was recorded in the Dover/Folkestone/Ashford triangle.

1928

6–7 January *Gale; storm surge; flooding*
A rare combination of events conspired to produce a severe flood on the Thames in the heart of London. Following a week of heavy rain and the thawing of the huge late December snowfall, there was extensive flooding the length of the Thames Valley, and the peak of the flood reached London on the 6th. A severe gale, initially south-westerly but veering north-westerly later, swept the country on the same date, with gusts to 84mph at Spurn Head, 83mph at Southport and 82mph at Fleetwood. This gale resulted in a substantial storm surge which travelled down the North Sea, almost coinciding with a high spring tide as it reached the Thames Estuary where the surge added an extra 2m to the natural tide. Districts bordering the Thames were flooded as far upstream as Hammersmith. Hundreds were made temporarily homeless, and fourteen people drowned in London's worst such flood probably since 1869.

10–11 February   *Severe gale*
A severe gale caused much damage to buildings and woodland in northern England, Wales, and the Midlands. Holyhead recorded a gust of 86mph, and a non-standard anemometer at Bidston Observatory on the Wirral Peninsula recorded one of 104mph.

28 June   *Flooding*
Flooding on the rivers draining Snowdonia followed a massive fall of rain which amounted to 197mm at Blaenau Ffestiniog. Over 100mm also fell in parts of the Lake District.

20 August   *Flooding*
A short-lived but severe flood hit Teesdale following a major downpour over the Yorkshire Dales. Barningham Park, North Yorkshire, collected 109mm of rain, and Kirkby Stephen in Cumbria 110mm.

22 October   *Tornado*
A short-lived tornado travelled from Victoria, through the West End, to Euston, causing a good deal of damage as well as consternation among thousands of people out for the evening. There were many reports of masonry falling into the streets, but no one was killed.

23 and 25 November   *Severe gales*
Two south-westerly gales in quick succession swept much of England and Wales, causing some loss of life and much disruption and damage. Gusts reached 85–90 mph at several sites on both dates, and 93mph at an over-exposed anemometer at Cardington, near Bedford.

1929

February   *Intense cold; heavy snow*
This month was mostly dry but very cold, and a period of intense cold lasted from the 10th to the 20th with the temperature continuously below zero in parts of eastern England and below –6°C on some days. At Ross-on-Wye the temperature fell to –18.3°C overnight 13th/14th. A large number of deaths from 'the cold', or

more specifically from exposure, were reported in the press. The month was also notable for one extraordinary but localized snowfall, when snow fell heavily for fifteen hours on the south-eastern flank of Dartmoor, where falls of 50–80cm were reported above Buckfastleigh and Ivybridge. The oft-quoted fall of 180cm at Holne Chase was in fact noted in a remote and unpopulated valley high on the moor above Huntingdon Cross, and may have been contributed to by earlier falls, and perhaps also by limited drifting from the high moor (although at lower levels the snowfall was remarkable for the complete absence of wind). There were widespread domestic consequences of the low temperatures, with many reports of exploding boilers as well as the more usual burst pipes; river traffic was suspended for several weeks, some harbours froze over, many roads in the West Country, Wales and western Scotland were blocked for a fortnight, and government figures showed that more than 150,000 men were put on the unemployment register as a direct result of the severe weather.

3–4 July *Flooding*
Floods hit Walton-on-the-Naze, Felixstowe, Ipswich, Stowmarket and Beccles following an overnight deluge lasting eight hours which dropped 115mm of rain at Rumburgh, near Beccles.

5 October *Flooding*
Floods affected parts of Cornwall, Devon and south-east Wales following prolonged heavy rain; Liskeard recorded 115mm, Princetown and Chagford 112mm and Abergavenny 109mm.

11 November to 9 December *Destructive floods*
This exceptionally wet period triggered river flooding in many parts of the UK, but by far the most damaging events occurred in the Welsh Valleys where torrents of water poured down from the hills on five successive Mondays, inundating more than 700 houses, tearing up roads and demolishing walls. On the 11th, 211mm fell at Lluest Wen reservoir in the upper Rhondda valley, on the 18th/19th, 178mm was recorded at Castell Nos reservoir, and there were further falls of around 100mm on the 24th–25th, 75mm on the 1st–2nd, and 125mm on the 6th–8th.

5–9 December   *Severe gales*
Considerable loss of life on land and at sea, as well as widespread
structural damage, accompanied five days of violent gales which hit
south-west England particularly severely. Highest recorded gusts
included 94mph at Pendennis Castle (Falmouth) on the 5th,
111mph at St Mary's (Scilly) on the 6th, 102mph at both Pendennis
and St Mary's on the 7th, and on the 8th and 9th 98mph and 87mph
respectively, both also at Pendennis.

1930

30 December 1929 to 15 January 1930   *Severe gales*
Gales returned on 30 December when a gust of 83mph was recorded
at Pendennis Castle, but the most serious repercussions followed a
violent gale which struck southern England and south Wales on the
12th, widely considered to be the most severe and most damaging
gale for half a century. The death toll may have exceeded 50, and
included the 23 crew of the Admiralty tug, *Saint Genny*, which went
down off Ushant. There was also extensive damage to buildings and
woodlands. Maximum gusts included 102mph at Pendennis and
97mph at St Mary's. Perhaps more remarkable was the gust of
94mph and the hourly mean of 70mph at Larkhill, an inland site in
Wiltshire.

17–18 June   *Thunderstorms; floods*
London was badly hit on both days, but the heaviest rain fell in the
northern parts of both Nottinghamshire and Derbyshire, with
104mm at Glossop. Ascot racecourse was flooded, and there were
several deaths from lightning. A cloudburst on Stainmore caused
much damage.

20–23 July   *Flooding*
An extraordinarily prolonged downpour over the North York
Moors triggered severe floods, especially on the Derwent, Esk and
Leven rivers. Many bridges were demolished, road and rail traffic
was seriously disrupted, telephone and telegraph services were suspended, and even the water supply was interrupted. Floodwaters
were up to 10m deep in places, and a sea-going vessel was used to

rescue two old ladies in Ruswarp, 3km inland. At Castleton, 145mm fell on the 22nd alone, with a phenomenal total of 304mm in four days.

6 August *Thunderstorms; floods*
Widespread thunderstorms were most severe in Somerset where 111mm fell at Cheddar.

17 September *Flooding*
Prolonged heavy rain over the Isle of Man led to flooding in many parts of the island, especially the northern half, and at Ramsey 110mm of rain fell during the day.

1931–40

1931

1–14 March *Heavy snow*
This was an exceptionally cold and very snowy spell for so late in the season. The heaviest falls occurred in northern and eastern Scotland on the 1st and 2nd, with level snow 45cm deep on Orkney.

27 May *Thunderstorms; heavy hail; flooding*
Severe thunderstorms broke out widely across western England and Wales; worst affected counties were Devon, Somerset, Glamorgan and Monmouthshire. Noteworthy rainfall totals included 114mm at Penarth, 113mm in Cardiff, 108mm at Abergavenny and 100mm at Bampton in Devon. In Cardiff the Taff and Ely rivers overtopped their banks, rail traffic was interrupted for three days and one person drowned in the River Wye. Many places reported heavy hail, notably at Watchet, Somerset, where stones with a diameter of 4cm were seen.

14 June *Tornado; thunderstorms; flooding*
One of the most destructive tornadoes of the twentieth century tore a 20km-long swath through Birmingham, from Small Heath to Erdington. There was a great deal of damage to homes and factories, one person died and several were injured. Elsewhere, serious

flooding was reported from Wick, Perth, Barrow-in-Furness and Cannock. A flash flood caused much damage at Bootle, Cumbria.

4–8 August   *Thunderstorms, flooding*
Severe storms accompanied by locally torrential rain occurred repeatedly during the month, but especially during these five days. On the 4th, 114mm of rain fell in 135 minutes at Steeple Langford in Wiltshire; on the 5th, 85mm fell in less than an hour at Chingford Mill in Essex; and on the 8th, 145mm was recorded, mostly within 150 minutes, at Boston in Lincolnshire where a large number of houses were flooded to a depth of 1m or more.

24 August   *Flooding*
A prolonged downpour led to flooding in the Channel Islands; St Helier recorded 115mm.

4 September   *Flooding*
For the second summer in succession a torrential downpour over the North York Moors led to severe flooding on the Derwent and Esk; on this occasion the Whitby lifeboat was used to rescue the same two ladies from their house in Ruswarp. Rainfall at Castleton amounted to 127mm. There were also floods in Leeds, and on the Derbyshire Derwent.

3–4 November   *Flooding*
Severe floods affected a large number of rivers, notably in south and mid-Wales, Devon and the Lake District. Over two days, 244mm of rain was measured at Trecastle in Breconshire, most of it falling in less than 36 hours. Over 200mm was also recorded at several sites above the Glamorgan/Monmouthshire coalfield, with at least 150mm over Dartmoor and the Lake District. On the 3rd alone, 100mm or more fell over a total area of 3,530 sq. km.

11 November   *Gales; storm surge*
A south-south-westerly gale produced a major storm surge in the English Channel which coincided with a high spring tide, resulting in widespread inundation of coastal districts, notably in Sussex, Hampshire and the Isle of Wight.

1932

2 to 14 January   *Severe gales; floods*
Damage occurred widely in southern England during the gale of
5th/6th, and in Scotland during that of 13th/14th. Highest gusts
included 92mph at St Mary's, Scilly on the 6th. Heavy rain on the
3rd triggered a severe flood in the catchment of the Irvine river;
Kilmarnock was badly affected, with water 2m deep in the town
centre. At Fenwick, Ayrshire, 100mm of rain fell.

1 May   *Thunderstorm; flooding*
Widespread but short-lived flooding affected Wiltshire, Oxfordshire,
Gloucestershire, Worcestershire, Herefordshire and Shropshire after
fierce thunderstorms broke out. Worst hit were Chippenham, and
also Calne (both Wiltshire) where 100mm of rain in estimated to
have fallen.

20–24 May   *Severe floods*
Destructive floods hit many parts of the north Midlands, Yorkshire
and Greater Manchester following prolonged heavy rain. Worst hit
were Birmingham, Coventry, Leamington, Bedworth, Nuneaton,
Burton-upon-Trent, Nottingham, Sheffield, Rotherham and Doncaster. At Bentley, near Doncaster, water stood in the streets for 35
days, while several coal mines were inundated and miners were laid
off. At Ecclesfield, just north of Sheffield, 101mm of rain fell on the
21st alone, with over 150mm between the 20th and 23rd. More than
75mm of rain fell over an area of 1,350 sq. km.

11 and 13 July   *Thunderstorms; flooding*
On the 11th, 131mm of rain was recorded at Cranwell, Lincolnshire,
most of it falling in 150 minutes. Heavy rain also hit the southern
Pennines, with 135mm at Barnoldswick (Lancashire) and 120mm at
Harewood Lodge (Cheshire). Cranwell was visited again two days
later when 61mm fell.

19 August   *Great heat and humidity*
The temperature reached 36°C widely in London and the Home
Counties, and with humidity levels also abnormally high, several
people died from heatstroke.

16–17 December   *Flooding*
In Borrowdale 234mm of rain fell, mostly within 36 hours. The Derwent catchment suffered most from flooding, and Derwentwater and Bassenthwaite Lake were effectively joined.

1933

23–26 February   *Snowstorm; gales; floods*
Although less disruptive than the Boxing Day snowstorm of 1927 in southern England, this was certainly one of the most extreme blizzards of the century in Wales, the Midlands and northern England. Level snow lay 70cm deep at Buxton and 76cm deep at Huddersfield, but the gale-force winds piled it up into drifts 4 to 5m high; gusts over 80mph were recorded along the English Channel coast. The snow was soft and clingy, telegraph poles and lines were brought down and thousands of trees were damaged. Transport throughout England and Wales was paralysed for three days. In southern England the snow soon turned to rain, and with three-day totals of 120mm at Selborne and 110mm at Petersfield (both in Hampshire), there was extensive flooding.

20–26 June   *Thunderstorms; hail; flooding*
This very thundery period produced no very large rainfall totals, but there were some exceptional falls in short periods, such as 64mm in 25 minutes at Silchester, Hampshire on the 24th. There was also catastrophic flooding on Merseyside on the 20th, notably in Birkenhead, Wallasey, Bootle and Litherland, where water was up to 3m deep; at least one person drowned.

26 September   *Thunderstorms; flooding*
Very localized but very intense storms struck parts of Hampshire and Wiltshire. Chippenham collected 109mm of rain in two hours, and Fleet reported 130mm in four hours.

9–10 October   *Severe flooding*
Destructive floods swept through the Welsh Valleys, just four years after the flooding disaster of 1929. In the Upper Rhondda Valley 194mm of rain fell inside 36 hours.

Prolonged drought
Water shortages, dried-out pasture and poor crop yields were the
result of an extended drought which lasted 34 months, from
November 1932 to August 1935, although the most intense phase
ended in November 1934. Three successive hot and sunny summers
exacerbated conditions. Averaged over England and Wales, the 25-
month intense phase produced 24 per cent less rain than usual, and
in parts of the south and east Midlands, south-east England, the
Thames Valley and Wessex, the shortfall amounted to 35 per cent.

1934

14 January   *Severe gales*
January and early February produced several gales, none worse than
that of the 14th which swept southern England and south Wales,
causing widespread damage and some loss of life. The highest gust
was 95mph at Pendennis Castle, Cornwall.

12 July   *Thunderstorms; flooding*
Flooding was reported in counties as widely spread as Glamorgan,
Suffolk and Cheshire. Highest rainfall totals for the day were
100mm at Ebbw Vale and 98mm at Wrexham.

22 July   *Severe thunderstorm*
Torrential rain, accompanied by hail, led to flooding in north-west
Kent and north-east Surrey; 116mm of rain fell in 100 minutes at
West Wickham, near Bromley.

8–10 October   *Flooding*
Prolonged heavy rain led to extensive flooding in Kent and Sussex.
Over the three days 125mm of rain fell at Hassocks and 114mm at
Keymer, both in mid-Sussex.

November   *Smog*
Two serious smog episodes occurred in this quiet, foggy month; one
lasted four days on Clydeside from the 12th to the 15th, while
London was smog-bound from the 19th to the 21st.

1935

24–26 January   *Severe gale*
High winds with damaging gusts swept Scotland, Northern Ireland and northern England; peak gusts included 100mph at Butt of Lewis, and 87mph at both South Shields and Bidston.

12–19 May   *Damaging frost; snow*
This spell of frost was the most severe late spring frost of the century, causing extensive damage to vegetable and soft fruit crops, and to apple and pear blossom, as well as to gardens and woodland in general. Up to 15cm of snow fell in several areas too. Air temperature dropped widely to between −3 and −6°C, and to −8.3°C at Rickmansworth, Hertfordshire.

23–25 June   *Thunderstorms, flooding*
On the 23rd, 141mm of rain fell at Llandwrog, south of Caernarfon, 108mm at Aberfeldy, Perthshire, and it is estimated that 200mm may have fallen above Auchnafree. On the 25th, 153mm was recorded in less than four hours at Swainswick, Somerset, and flooding affected hundreds of homes and businesses in Bath.

16–17 September   *Severe gale*
Highest gusts included 98mph at Pendennis Castle, 96mph at St Mary's and over 80mph at several inland sites in southern England. There was some loss of life and extensive damage.

22 September   *Hailstorm*
A narrow zone stretching 65km from Banbury (Oxfordshire) to Irthlingborough (Northamptonshire) was hit by one of the century's most dramatic hailstorms; Peterborough was also affected. Glasshouses, roofs and outhouses were destroyed. Stones 10cm across were picked up at Great Billing, near Northampton.

19 October   *Severe gale*
The SS *Vardulia* was lost with 37 hands off the Western Isles, there were other casualties at sea and eleven deaths on land, putting the overall death toll at over 50. Structural damage was widely reported

across the northern half of the UK. Maximum gust was 92mph at Renfrew aerodrome.

18–23 December *Smog*
Widespread fog persisted for long periods during this week over many parts of the UK, and pollution levels rose to their highest for over ten years in London and other conurbations.

1936

9–10 January *Severe gales*
High winds caused damage to buildings and woodland in many parts of the country; highest gusts were 92mph at Bidston on the Wirral and 91mph at the Lizard in Cornwall.

19–20 January *Heavy snow*
Widespread snow, heaviest from the north Midlands northwards, blocked roads and railways and disrupted telephone and telegraph services. Some main roads on Speyside were blocked for ten days. Snow lay 45cm deep at Buxton and 30cm at Bellingham (Northumberland).

May to August *Thunderstorms; hailstorms; flooding*
Summer 1936 was exceptionally thundery with frequent hail and a good deal of local flooding. A hailstorm – stones 4cm across – caused much damage in the Dunstable district on 6 May and 53mm of rain fell in 45 minutes at nearby Houghton Regis. Torrential downpours triggered flooding on 21 June at St Albans where 112mm of rain was recorded, and Nuneaton which had 102mm. Severe urban flooding was noted in Bristol on 28 June.

25–27 October *Severe gale*
There was much damage in Scotland and northern England thanks to this ferocious storm; gusts included 104mph at Tiree and 95mph at Paisley.

19–28 November *Smog*
Heavily polluted fog lasted without a break from the 21st to the 26th

in Manchester, and for long periods in Liverpool and Birmingham, and caused considerable difficulty to the transport system. Cotton mills were closed in the Manchester area because the pollution damaged the textiles.

3–18 December *Severe gales; extensive flooding; freezing rain*
A sequence of severe gales swept the country; they were most violent in Scotland around the 3rd to the 6th and the 16th to the 18th, with gusts to 99mph at Bell Rock during the first period and to 95mph at Stornoway during the second. Widespread floods affected Wales, north-west England and Scotland mid-month, and Lancaster was completely cut off for a time. At Trecastle in Breconshire 117mm of rain fell on the 13th alone, and 256mm between the 13th and 17th; on the 19th, 147mm fell in Glen Shiel, 137mm in Glen Leven, and 127mm at Fort William. Many road accidents followed widespread freezing rain on the 8th in the Midlands and south-east.

1937

29 January *Heavy snow*
Heavy falls were confined to north and east Scotland: 37cm in Aberdeenshire, 25cm in Fife.

27–28 February *Snowstorm; severe gale*
A fierce blizzard raged over much of the country, and worst hit were the north Midlands, north Wales, all of northern England, and southern and eastern Scotland. Level snow was 60cm deep at Buxton, and 25–50cm over wide areas, but the northerly gale swept the snow into drifts 4 to 5m high. Roads and railways were blocked, not only by drifts but by trees and telegraph poles brought down in the strong winds. Power and telephone lines were damaged and the BBC was forced off the air in some regions. A gust of 107mph was recorded at Holyhead.

11–13 March *Heavy snow*
Worst hit were northern England, southern Scotland and Northern Ireland. In the province all roads were closed for a time, snow lay 25–35cm deep with drifts to 3m and five deaths were reported. Over 60cm of snow fell at Barnard Castle in County Durham.

January to March *Flooding*
With rainfall 65 per cent above normal over the three-month period, this represents the wettest start to a year over England and Wales in 300 years of rainfall records. There were very few individual falls of note, but the incessant rains led to widespread flooding which culminated in a disastrous inundation of large parts of the Fen District which peaked on 17 March.

15 July *Thunderstorms; flooding*
Lightning, hail and flood damage occurred in many parts of the country, with Leicestershire, Lincolnshire, Dorset, Somerset and Gloucestershire worst hit. Highest rainfall totals for the day included 146mm at Waltham-on-the-Wolds (Leicestershire), 139mm at Wyberton and Boston (both Lincolnshire), 116mm at Belvoir Castle (Leicestershire) and 106mm at Pensford (Somerset). Four days later another thunderstorm caused serious flooding in Spalding, with 54mm falling in 25 minutes.

7 December *Flooding*
Over 50mm of rain fell over most of the Isle of Wight, with 105mm at Newport. The flooding on the island was described as the worst for over 40 years.

11–15 December *Heavy snow*
Northern England and Scotland were badly hit by this snowstorm; 45cm fell at Balmoral, 35cm at Bellingham and 28cm at Durham observatory.

1938

January–February *Frequent gales; storm surge*
Severe south-westerly or westerly gales swept large parts of the UK on 14–15 January when a gust of 101mph was recorded at St Anne's Head, Pembrokeshire, and another on 29 January when Bidston (on the Wirral) also recorded a gust of 101 mph. On both occasions a storm surge coincided with spring tides to bring coastal flooding to the Bristol Channel, Cardigan Bay, Lancashire and Cumbria. Five crew-members drowned when the *Alba* capsized off Cornwall. A

northerly gale triggered a storm surge in the North Sea on 11–12 February, again coinciding with high spring tides, and serious flooding affected eastern counties from Yorkshire to Kent.

1–2 June   *Severe gale*
A rare summer gale caused considerable damage in southern England and south Wales; the highest gusts were 88mph at Calshot, Hampshire and 80mph at the Lizard, Cornwall.

4 August   *Thunderstorms; flooding*
Tremendous storms battered south Devon with torrential rain, hail and lightning strikes. A total of 162mm of rain fell in under six hours at Torquay, with 108mm in 110 minutes. Other noteworthy rainfall totals were 164mm at Hedgebarton, 152mm at Ilsington and 152mm at Stoke Gabriel. More than 100mm of rain fell over an area of almost 500 sq. km.

11 August   *Thunderstorm; hail; flash flood*
Strathaven, some 20km south of Glasgow, was deluged with 130mm in little more than two hours. A 'waterspout' was observed nearby. Landscape damage was considerable.

4 October   *Severe gale*
Wales, the Midlands and northern England were worst hit; peak gusts included 104mph at St Anne's Head, 95mph at Bidston on the Wirral and 91mph at Barton Road aerodrome, Manchester.

23 November   *Severe gale*
Yet another violent gale battered England, Wales, Northern Ireland and southern Scotland, with highest gusts of 108mph at St Anne's Head and 91mph at the inland site of Mildenhall in Suffolk. Structural damage occurred widely and more than ten people lost their lives.

18–26 December   *Severe cold; heavy snow*
After one of the warmest autumns on record, intense cold set in with dramatic suddenness on the 17th/18th and the temperature remained below freezing over parts of south-east England for nine consecutive days. Snow fell daily, and was heaviest on the 21st–22nd

with 30–50cm falling over much of eastern England, and 60cm in Lincolnshire and west Durham.

At intervals throughout the year   *Heavy rain*
The year was remarkable for the repeated bouts of prolonged heavy rain in upland regions, especially in the Western Highlands, the Lake District and Snowdonia, and although these were not, strictly speaking, disasters, they still deserve a mention. Kinlochquoich in western Inverness-shire recorded 221mm of rain in 48 hours on 2–3 February and 201mm on 29–30 March; Hassness, above Buttermere, collected 234mm on 29–30 July; Watendlath Farm in Cumbria had 187mm on 2–3 October; and Kinlochquoich (again) measured 227mm on 5–6 November. In Argyllshire and Inverness-shire several places had between 70 and 90 per cent more rain than in an average year. There was repeated flooding on local rivers and burns, although relatively few people were directly affected. A further 201mm fell at Kinlochquoich on 9–10 February 1939.

1939

22–23 January   *Severe gale*
A powerful north-westerly gale affected chiefly Wales and south-west England during the night, the wind gusting to 96mph at St Mary's (Isles of Scilly) and to 91mph at the Lizard. Seven members of the St Ives' lifeboat crew were drowned going to the aid of an unknown vessel.

25–26 January   *Snowstorm; flooding*
Heavy snow fell across central and southern England; at low levels the snow thawed readily, but above about 100m it accumulated to depths of 30–60cm. Some 60cm fell over the Downs of Hampshire, Wiltshire and Berkshire, and over the Chiltern Hills where drifts reached 4m in height. Even here the snow was soft and clinging, and it brought down hundreds of telegraph poles and thousands of branches. There was a railway accident at Hatfield (Hertfordshire) and another train was stranded in drifts in south Wales. Flooding followed the snow; Suffolk was badly hit with 75mm of rain at

Hadleigh, and the floods in Ipswich were the worst of the entire century. Some 3,000 families were evacuated from their homes in the Manchester area.

20 August to 2 September *Thunderstorms; flooding*
This very thundery episode began on the 20th with a destructive flood in east Dorset, with 118mm of rain in six hours at Bryanston School and 115mm at Bryanston Gardens, both near Blandford Forum. The next day, Frome (Somerset), Oxford and London were badly hit, and seven people sheltering in Valentine's Park, Ilford, were killed by lighning. On the 29th, 102mm of rain was recorded at Milford, Staffordshire, and the nearby LMS railway line was flooded to a depth of over 1m. On 2 September, widespread thunderstorms announced the outbreak of war; Conisborough in Yorkshire reported 103mm of rain.

27–28 October *Flooding*
Severe flooding affected several towns in east Kent, not least Folkestone where 120mm fell.

1940

December 1939–February 1940 *Severe winter*
The first winter of the Second World War was the coldest since 1895, and the snowiest (in terms of persistence of snow on the ground) since 1879. Deaths from hypothermia were not routinely enumerated before 1982, but a conservative estimate for the winter of 1939–40 is between 1,000 and 2,000. The mercury fell to –23°C at Rhayader in mid-Wales on 21 January, and there were lengthy periods when the temperature remained below zero by day and by night.

26–30 January *Snowstorm; icestorm; floods*
A prolonged, heavy and widespread snowfall affected most parts of the UK, paralysing the transportation and communications networks. An already critical situation was exacerbated on the 28th by heavy freezing rain which affected a broad band stretching from Sussex and Hampshire across the southern and western Midlands to mid- and north Wales. Unusually for Britain, the freezing rain was

not immediately followed by a thaw, and the ice persisted in many places for six or seven days. Everything was coated with 2–3cm of clear ice, and the weight brought down a great many telegraph poles and wires, electricity pylons and lines, and the damage to trees was enormous. There were also many street accidents, and hospitals were inundated with pedestrians with broken bones. By the end of the month, snow lay 50–60cm deep over much of Yorkshire, Lancashire, the north Midlands, Kent and Sussex, and 38cm deep in central Birmingham; drifts to 4 or 5m were typical. A thaw which began on 4 February triggered widespread flooding.

10–26 July   *Thunderstorms; hailstorms; flash floods*
On the 10th, 106mm of rain was recorded in three hours at Osgodby, near Scarborough, while in Lincolnshire Horncastle collected 98mm and Grimsby 89mm. Large hailstones destroyed glasshouses and damaged windows and roofs in Boston and East Kirkby (both Lincolnshire), and on the 11th two children were killed by lightning at Horfield, near Bristol. Floods hit the Cleveland coast on the 17th, with 90mm of rain falling at Redcar, and on the 26th a destructive flash flood tore through the small town of Cromarty, north of Inverness, demolishing four houses and a bridge, after 64mm of rain fell in a few hours.

November   *Flooding*
One of the wettest Novembers in three centuries of records led to flooding along the Thames, Severn, Wye, Great Ouse and Trent. A fall of 240mm on the 2nd–3rd at Llyn Fawr reservoir, at the head of the Rhondda Valley, led to serious local flooding in the valleys of Glamorgan and Monmouthshire.

11–12 November   *Severe gale*
A severe gale swept England and Wales; a gust of 103mph was recorded at St Anne's Head.

6 December   *Severe gale*
High winds battered England and Wales; peak gusts included 96mph at St Anne's Head, 94mph at Bidston observatory, and 90 mph at the Lizard.

1941–50

1941

20–22 January   *Heavy snow*
Snow fell widely and heavily over most of the UK; on the 20th it was 38cm deep at Edgbaston, Birmingham; on the 21st it was 38cm at Lake Bala in north Wales; and on the 22nd it was 50cm at Balmoral, on Royal Deeside. A subsequent thaw over England and Wales caused flooding, notably along the Severn where the flood-level at Shrewsbury was the highest since 1881. Further snow in early February took the snow depth at Balmoral to 68cm.

18–20 February   *Snowstorm*
The great north-east snowstorm: an area stretching from the Humber to the Forth was badly hit by this massive snowfall. Most transport routes through the area ceased to function for several days, six trains (with a total of more than 1,000 passengers) were stranded in drifts between Newcastle and Edinburgh, and telegraph and power lines were brought down. Level snow depths included 122cm at Consett, 107cm at Durham, 76cm at Newcastle and 66cm at Marchmont in Berwickshire. Severe flooding followed a thaw.

27–31 March   *Heavy snow*
Heavy snowfalls hit northern Scotland; at Tain, Cromarty, it was reported to be the worst snowstorm for over 60 years with 80–90cm level snow on the ground.

4–15 May   *Severe frost*
Damaging frosts occurred widely over England and Wales on several nights; vegetable and fruit crops were ruined in the Vale of Evesham, East Anglia and Kent. Lowest temperature was −9.4°C at Lynford, Norfolk, on both the 4th and the 11th.

23 May to 26 July   *Thunderstorms; flooding*
Thunderstorms occurred frequently during this period. On 23 May, 109mm fell at Linlithgow (West Lothian), 104mm at Polmont (Stirlingshire) and 99mm at Morton (Midlothian), and an aqueduct was

washed away. On 12–13 June Upton Warren (Worcestershire) recorded 117mm in 25 hours, and two people in Selly Oak, Birmingham, were killed by lightning. On 22 June, Newcastle-upon-Tyne registered 113mm of rain and there was considerable hail damage in the district. Finally, on 26 July 115mm fell at Wickham Bishops and 108mm at Writtle (both Essex) leading to serious flooding in the Chelmsford, Witham and Maldon districts.

5 October   *Thunderstorms; flooding*
A heavy storm broke over an area more used to prolonged orographically enhanced rain; 130mm fell at Taf Fechan reservoir, above Merthyr Tydfil, mostly in six hours.

1942

19–20 January   *Snowstorms; flooding*
Snow fell over practically the entire country, but was heaviest in the west Midlands, northern England and southern Scotland. Edgbaston, Birmingham, had 30cm, West Linton, Peebles-shire, had 60cm, and drifts as high as 6m were reported from many parts of southern Scotland. A thaw later in the month led to extensive flooding in central and northern England, but further snow fell in Scotland. Several villages in north-east Scotland were isolated for over a week.

4–7 March   *Heavy snow*
Widespread snow, heaviest in northern England where Huddersfield reported 30cm.

30 June   *Thunderstorm; hail; flooding*
Violent but short-lived thunderstorms with large hail hit west and south London. At New Malden, Surrey, 55mm of rain fell in 33 minutes. Several lightning deaths were reported.

29–30 August   *Flooding*
Heavy rain triggered flooding in north Somerset and south Gloucestershire. Burnham-on-Sea in Somerset recorded 114mm of rain, of which 57mm fell in 50 minutes.

25 October   *Flooding*
Severe floods affected much of the Isle of Man. West Baldwin reservoir recorded 103mm.

1943

7 April   *Severe gale*
A destructive gale, all the more unusual for occurring in April and during the morning, battered in particular southern Scotland and northern England. Maximum gusts were 91mph at Eskdalemuir, Spurn Head and Ringway aerodrome, and 88mph at South Shields.

10–11 July   *Flooding*
A prolonged downpour over the Welsh Valleys triggered flooding; at Ogmore Vale, north of Bridgend, 151mm of rain fell in 48 hours.

3–5 October   *Flooding*
Flooding in the Central Lowlands followed heavy rain on the 3rd when 109mm fell at Castle Lachlan (Argyllshire), 108mm at Lochwinnoch (Renfrewshire) and 107mm at Lennoxtown (Stirlingshire). The situation was exacerbated by a further 60–80mm of rain on the 5th.

1944

22–25 January   *Severe gale; flooding*
Persistent heavy rain led to flooding, notably in Wales and in Yorkshire where the River Aire overtopped its banks in several places, including Leeds. On the 22nd alone, Blaenau Ffestiniog collected 119mm of rain, Machynlleth 118mm, and Great Hay in Yorkshire 87mm. Gales were particularly severe on the 24th when St Anne's Head had a gust of 94mph.

27 February to 2 March   *Snowstorms*
Snow fell widely on the 27th, with 20–30cm reported widely in the north Midlands, northern England and Scotland. Derry Lodge, above Braemar, reported 40cm, while drifts 10m high cut off towns and villages in the far north of Scotland.

29–30 May   *Thunderstorms*
A flash flood on the 29th caused much damage in Holmfirth in Yorkshire; 114mm of rain was recorded at Swineshaw, above Glossop, but 150–175mm is estimated to have fallen above Holmfirth. The next day 113mm fell at Kilve in Somerset, triggering severe local flooding. Hailstones 3cm across were reported in both storms. At least four people died.

5–6 June   *Gales*
This is the disaster that might have been. Strong winds and low cloud caused the postponement of the D-Day landings on the 5th, but a better forecast for the 6th allowed the operation to proceed. The legend persists, including naturally enough in meteorological circles, that the forecasters were heroes, having identified a 'weather window'. The truth is less romantic: they had very limited information to produce their forecasts, their prediction for the 6th was far too optimistic, and had they got it right the invasion would have been postponed for two weeks. Very strong north-easterly winds between 18 and 22 June would have meant either a disastrous second attempt, or, more likely, another postponement. In the event it was probably a good thing that the forecast for 6 June 1944 was wrong.

11 October   *Severe gale*
Southern England was swept by a gale; St Anne's Head reported a gust of 97mph and Pendennis one of 89mph.

16–17 November   *Flooding*
Rivers flooded in all southern counties from Kent to Cornwall, but Devon was worst hit. South Brent, on the southern fringe of Dartmoor, recorded 182mm of rain in 48 hours.

19–29 December   *Smog*
Thick fog cloaked much of England, Wales and southern Scotland, and was particularly persistent in urban areas where high pollution levels led to a filthy, choking smog.

1945

9–10 January   *Heavy snow*
Worst hit were eastern Scotland and north-east England; Belling-ham, Northumberland, reported 60cm of level snow.

18–19 January   *Severe gale*
Widespread structural and landscape damage accompanied this exceptional westerly gale which affected all parts of the UK. The wind gusted to 113mph at St Anne's Head, 89mph at St Mary's, Scilly and 88mph at South Shields.

25–30 January   *Snowstorms*
On the 25th south-west England, the Midlands and Wales were badly hit, with 35–45cm of snow falling widely and 75cm in the Cardiff area. On the 29th–30th north-east England and Scotland were paralysed by a blizzard which dropped 25–35cm over a large area. Drifts of 6m blocked roads and railways; several trains were trapped in north-east Scotland.

11 May   *Thunderstorms; hailstorms*
Fierce storms struck the south Midlands, hailstones 3–4cm across were widely reported from Leicestershire, Northamptonshire, Oxfordshire and Buckinghamshire.

14–15 July   *Thunderstorms; flooding*
Vivid lightning was a feature of these prolonged overnight storms, but there was flooding as well especially in Surrey, Sussex and Hampshire; 107mm of rain fell at Petworth.

28–29 August   *Thunderstorms, flooding*
Newport, Monmouthshire, suffered a serious flood on the 28th after 76mm of rain fell in less than two hours; on the 29th Masham Moor, Yorkshire, had 102mm and Boston (Lincolnshire) 97mm.

1946

8 January   *Severe gale*
Gale damage was confined to Scotland; Stornoway recorded a maximum gust of 99 mph.

7–8 February   *Flooding*
Following an exceptionally wet January, prolonged heavy rain affected much of the country, with over 150mm in 48 hours in mid- and north Wales. Extensive floods were reported on the Wye, Severn, Trent and several rivers in Wales and Cheshire.

20 February to 10 March   *Heavy snow*
Frequent snowfalls, heaviest in Scotland. Braemar had 45cm level snow with 2m drifts.

2–3 July   *Thunderstorms; flooding*
Westley Hall, near Bury St Edmunds, endured deluges on two successive days, totalling 130mm. Flooding in Bury St Edmunds itself was serious, but short-lived.

26 July   *Thunderstorms; flooding*
Kew Observatory recorded 50mm in 35 minutes, and 80–100mm of rain fell widely in East Anglia. The floods in east Norfolk were described as the worst since August 1912.

18–20 September   *Severe gale; flooding*
Holmfirth in Yorkshire had 90mm on the 19th, and Blaenau Ffestiniog 253mm in 72 hours. Devastating floods affected large parts of Wales, Lancashire and Yorkshire; Bradford and district were dealt a particularly harsh blow. A gust of 100mph was recorded at St Mary's.

23 November   *Flooding*
Princetown, on Dartmoor, collected 173mm of rain, and Castell Nos reservoir, in the upper Rhondda Valley, had 156mm. Floods hit the West Country, south Wales and also the Thames.

1947

January to March   *Severe winter; snowstorms*
The most famous winter of the twentieth century – the coldest first
quarter since 1838, the snowiest since at least 1814 and possibly
since 1684 – combined with post-war privations, fuel shortages and
industrial unrest to bring Britain to its knees. The cold and snowy
weather lasted from 22 January to 14 March in the southern half of
the UK, and from 24 January to 17 March in the northern half.
Major snowfalls, each contributing 25cm or more over a wide area,
occurred on 23 January, 28 January, 2–4 February, 8–9 February,
25–26 February, 4–5 March, 8–9 March and 12–13 March. Accumulated snow depths peaked at 41cm at Edgbaston in Birmingham,
44cm at Harrogate, 53cm at Wrexham, 64cm at Huddersfield, 86cm
at Ushaw near Durham, 92cm at Lake Vyrnwy, 135cm at Forrest-in-
Teesdale in the south-western part of County Durham and 152cm
at Clawdd Newydd, near Ruthin, in north-east Wales. Drifts 8m
high were reported from the Scottish Borders in mid-March. The
coldest episode was between 11 and 23 February when the temperature remained continuously below zero over a large part of the UK.
For good measure there was a severe freezing-rain event across
southern counties of England on 5 March. The effects on agriculture were very serious: cereal yields in 1947 were 10–20 per cent
below their 1946 levels, and did not fully recover until 1949; some
20–25 per cent of the UK's sheep were lost, and the 1946 total was
not reached again until 1952.

Second half of March   *Thaw floods; severe gale; storm surge*
A destructive gale swept central and southern parts of England and
Wales on 16 March, with gusts of 98mph at Mildenhall, Suffolk, and
93mph at Cardington, Bedfordshire. Repeated bouts of heavy rain
conspired with a rapid thaw of the deep snow-cover to create inland
floods which were unprecedented in their severity, longevity and
geographical extent. All of the major river catchments of England
and Wales were badly affected, though no region suffered more than
the Fen District. A North Sea storm surge on the 16th triggered
coastal flooding as well.

23 April   *Severe gale*
A further damaging gale hit much of Britain with gusts of 88mph at Aberporth (Ceredigion) and over 80mph widely across the interior of England and Wales.

16–18 July   *Thunderstorm; hail; flooding*
Serious damage to the Royal Horticultural Society Gardens at Wisley, Surrey, was caused by a downpour of 128mm on the 16th, most of which fell in less than 75 minutes. On the 17th, 91mm fell in two hours at Wittersham, Kent, and on the 18th Woodstock in Oxfordshire recorded 89mm in 105 minutes. Railway services in the Manchester area were suspended on the 16th following the accumulation of deep floodwaters.

20–21 November   *Flooding*
A prolonged downpour conspired with a rapid thaw to cause the worst flood for decades in western Scotland with bridges demolished, roads washed away and stock drowned. Glenquoich recorded 210mm and Glenshiel 194mm, both in 48 hours.

13 December   *Storm surge*
Strong winds and high tides caused flooding along the Channel coast, notably at Portland.

1948

January   *Flooding*
This was the wettest January of the century, and on Wearside and Tyneside more than four times the normal amount of rain fell. Long-lasting floodplain inundations affected most of Britain's river systems, and many homes were flooded too. The level of the Severn was 6m above normal at Shrewsbury and 4.5m above at Worcester.

8–9 February   *Severe gale*
Much damage occurred over central and southern Scotland, northern England, north Wales and Northern Ireland. Highest gusts included 95mph at Durham and 87mph at Renfrew.

21–22 February   *Heavy snow*
Dry powdery snow fell across south-east England; Biggin Hill recorded 36cm with 2m drifts.

18 July to 8 August   *Thunderstorms; flooding*
A protracted period of thundery weather with torrential local downpours, including 102mm at Neath (Glamorgan) and 100mm at Silsoe (Bedfordshire) on the 2nd, and 108mm at Stalbridge (Dorset) on the 8th, most of which fell inside two hours.

11–12 August   *Severe flooding*
The great Border flood: a prolonged downpour triggered floods in the north Midlands, north-east England, the Scottish Borders and the Lothians, but these floods were catastrophic only in the Tweed catchment where 158mm fell on the 12th alone at Kelso (172mm in 48 hours), 141mm at Paxton and 139mm at Kingside. The weekly total (6th–12th) at Kingside was 270mm. On the 12th, 100mm or more fell over an area of 2,100 sq. km. Forty bridges were washed away, the east coast main railway line was out of commission for three months, thousands of livestock were lost, and arable land was devastated.

22 November to 1 December   *Smog*
Often forgotten in the wake of the 1952 smog, this was a very serious episode with over 1,000 deaths attributed to high pollution levels. At Kingsway, in central London, the fog persisted without a break for 113 hours, from 9pm on the 26th to 2pm on the 1st. It was particularly dense on the 27th when severe traffic dislocation occurred over much of the UK.

1949

15–17 July   *Thunderstorms; flooding*
The heaviest downpours during this brief thundery period occurred on the 15th – notably a fall of 105mm at March (Cambridgeshire) most of which fell in 105 minutes. Lightning killed several people on the 16th in London and the Home Counties, and severe short-lived flooding accompanied downpours in Somerset on the 17th.

25 October   *Flooding*
At Linhope, in the Northumberland Cheviots, 122mm of rain fell, triggering serious flooding.

1950

2 February   *Severe gale*
A damaging gale affected much of England and Wales except Cumbria and Northumberland, and south-west England was worst hit with gusts to 104mph at Pendennis Point, near Falmouth, and 92mph at Mountbatten in Plymouth.

February   *Flooding*
One of the wettest Februarys on record led to widespread flooding on rivers in England and Wales during the second week, while Scotland was badly hit between the 16th and 19th as heavy rain and a rapid thaw conspired.

10 April   *Severe gale*
Gale damage was reported widely from northern England; maximum gusts included 85 mph at Fleetwood and 83mph at Southport.

25–26 April   *Heavy snow*
An exceptional late snowstorm hit southern England, with 15–20cm falling, mostly within six hours, above the 100m contour from Wiltshire to Kent. The snow was soft and clinging, causing damage to trees, and bringing down hundreds of kilometres of telephone lines.

21 May   *Tornado; hail; heavy rain*
Britain's longest-lasting tornado tracked from Wendover, Buckinghamshire, to Ely, Cambridgeshire; the track was 105km long, and the tornado lasted 150 minutes. Much damage was done in Linslade and the southern fringes of Bedford, while a double-decker bus was overturned at Ely. Violent thunderstorms accompanied by 5cm hailstones were also reported close to the track. No deaths were reported.

6 September   *Flooding; gales*
A wild day, the wind gusting to 81mph at Spurn Head, while
115mm of rain fell at Coniston and 108mm at Ulpha, both in the
Lake District. Flooding was widely reported, including in urban
areas in Scotland, Northern Ireland and northern England. The
downpour contributed to the Knockshinnock Colliery disaster in
Ayrshire where thirteen miners were killed.

3–6 December   *Heavy snow*
A 20–30cm blanket of snow covered a large area from the north
Midlands to northern Scotland, with 51cm measured at Dalwhin-
nie, Inverness-shire.

15–17 December   *Heavy snow*
Snow fell heavily over much of the UK, with greatest depths
reported from eastern and southern coastal counties. Scarborough
and Lowestoft had 36cm, and large parts of Dorset, Hampshire and
the Isle of Wight 25–30cm.

1951–60

1951

1–10 January   *Heavy snow*
The severe December weather continued for ten more days: during
the opening days Buxton had 31cm of level snow and Whipsnade
in Bedfordshire 30cm; on the 9th, 30cm lay over large parts of
Scotland. Heavy rain and a rapid thaw led to extensive flood-plain
inundation in late January and February.

10–11 April   *Flooding*
Following the wet winter, there was renewed flooding on the Great
Ouse, the Thames and in Norfolk where the Broads were badly hit.

22–31 July   *Thunderstorms; hail; flooding*
A very thundery period with short-lived torrential downpours,
notably 76mm in two hours at Farnham on the 22nd, and 62mm in

45 minutes at Cowes on the 30th. Severe local floods were reported from Belfast, Aberdeen, Dundee, Scarborough, Chester, Rochester, Bridgwater and Brighton; large hail was reported from many places, and at least three people were killed by lightning.

4 November   *Severe flooding*
Perthshire, Angus, Kincardineshire, Aberdeenshire, Argyllshire and Inverness-shire were badly affected by flooding; 111mm fell at Ardeonaig Manse and 103mm at Glen Ogle (both Perthshire).

30 December   *Severe gale*
A severe gale swept Scotland and northern England; it was particularly destructive in the Central Lowlands where Millport recorded a gust of 108mph.

1952

15 January   *Severe gale*
Known as the 'Orkney hurricane', this gale battered central and northern Scotland, and caused widespread destruction and some loss of life in Caithness, Sutherland, Lewis, Orkney and Shetland. Peak gusts at official stations included 120mph at Kirkwall, 108mph at Stornoway, 104mph at Lerwick and 100mph at Wick. An anemometer at the Electrical Research Association site at Costa Hill, a very exposed site near the northern extremity of the main island of Orkney, registered a gust of at least 126mph.

29–30 March   *Snowstorm*
Heavy snow whipped into huge drifts by winds gusting to 60–70mph fell across much of southern and central England and south Wales. At Whipsnade, Bedfordshire, level snow lay 20–25cm deep with drifts to 2.5m.

19 May   *Thunderstorms; hail; tornado*
Violent thunderstorms triggered severe, if short-lived, flooding in Devon, Oxfordshire, Berkshire and Surrey. Large hail was also reported. 73mm of rain fell at Winnersh, near Reading. A tornado caused a good deal of damage in Tibshelf, Derbyshire.

6–9 August   *Thunderstorms; flooding*
Exceptional downpour on the 6th in north-west London and south Hertfordshire: 123mm, mostly in two hours, at Borehamwood. Severe storms in central Pennines on the 9th: 105mm at Slaidburn.

15–16 August   *Flash flood*
The Lynmouth disaster: a flash flood tore through the small coastal town of Lynmouth, killing 34 people, making 420 homeless, demolishing 93 buildings and 28 bridges, and carrying 38 motor cars out to sea. Rainfall totals, mostly in less than seven hours, included 229mm at Longstone Barrow, 192mm at Challacombe and 187mm at Simonsbath. More than 100mm fell over an area of 640 sq. km. An estimated 280–300mm fell just north of Simonsbath.

15 August   *Flooding*
The Scottish River Dee flooded badly following a fall of 104mm at Birkhall, near Ballater, on Deeside.

6 November   *Severe gale*
A ferocious north-westerly gale affected much of the UK, causing widespread damage in north-west England, Wales and the Midlands. Highest gusts were 97mph at Bidston on the Wirral, 94mph at Shawbury and 85mph at Fleetwood.

5–9 December   *Smog*
The great London smog: thick fog descended upon a large part of southern and central England, but in London it was heavily laden with sulphur dioxide, nitrogen oxides and particulate matter. At Kingsway in central London the fog lasted for 114 hours. Most references use government figures to put the death toll at 4,000, but that analysis adopted an extremely conservative mortality time-frame and counts casualties only in the old London County Council area. Other estimates suggest that between 7,000 and 12,000 lost their lives.

14–18 December   *Heavy snow; severe gale*
Heavy snow affected central and northern districts on the 14th and 15th, with 41cm at Bwlchgwyn in Denbighshire and 31cm at West Kirby on the Wirral in Cheshire, while massive drifts in western and

northern Scotland led to many communities being cut off for several days. A violent gale followed on the 17th–18th, with gusts to 110mph at Cranwell (Lincolnshire), 92mph at Stornoway, 91mph at Tiree and 91mph also at Fleetwood.

1953

31 January to 1 February   *Severe gale; storm surge; severe coastal flooding*
The North Sea flood: a violent north-westerly gale conspired with high spring tides to bring about the most extensive coastal flooding in the UK probably for three centuries. The North Sea storm surge amounted to 2.0–2.5m between the Humber and Thames estuaries, and in total some 6,500 sq. km of land was inundated. The UK death toll was 307, including 132 who drowned when the ferry *Princess Victoria* went down in the North Channel on the afternoon of the 31st. Several thousand families were evacuated from their homes. In Scotland the gale flattened hundreds of hectares of timber, and widespread structural damage occurred in the north and east, including Aberdeen. At Costa Hill, Orkney, a mean wind speed of 90mph lasted for almost five hours, with gusts to 125mph; elsewhere, Kirkwall recorded gusts to 108mph, Aberdeen airport to 101mph and Lerwick to 98 mph.

8–14 February   *Heavy snow*
Heavy snow fell in Wales, the Midlands, all of northern England and the eastern half of Scotland. It was reported to be 60–90cm deep at Malham Tarn, with 43cm at Bwlchgwyn (Denbighshire), 36cm at Buxton and 32cm at Lake Vyrnwy.

25 May   *Severe thunderstorms*
Widespread thunder and hailstorms occurred, and at least four people were killed by lightning. Severe floods affected Ballykelly in County Londonderry, and Fort William.

16–26 June   *Severe thunderstorms*
On the 16th, 106mm of rain fell at Sedbergh (North Yorkshire) with flooding in the town and further down the Lune Valley. On the 26th

a violent downpour hit Eskdalemuir (Dumfriesshire), washing away bridges and flooding the village; a total of 107mm fell, of which 80mm came down in 30 minutes. Elsewhere, 92mm was recorded at Malham Tarn in the Yorkshire Dales, and 44mm in fifteen minutes at Nelson, Lancashire, where hundreds of homes were inundated. The Kirkstone Pass was blocked by boulders after a rockslide.

1954

15 January   *Severe gale*
Structural damage in Scotland and northern England; several deaths; gusts to 97mph at Renfrew, 94mph at Millport and 91mph at Flamborough Head.

20 January   *Flooding*
Many parts of Lancashire were flooded; 104mm fell at Slaidburn and 103mm at Abbeystead.

25 January to 9 February   *Severe cold; heavy snow*
Intensely cold with the temperature continuously below freezing for over nine days in parts of southern England. Heavy snowfalls included 36cm at Clun (Shropshire) and 31cm at Evancoyd (Radnorshire). Drifts of 2–2.5m isolated many villages in Kent.

5 June   *Thunderstorm; flooding*
At Wyton, near Huntingdon, 90mm fell in a three-hour storm, of which 80mm came down in 105 minutes. Severe local flooding ensued.

May to September   *Poor summer*
Generally accepted as the poorest summer in the second half of the century, it was particularly notable for an almost complete absence of warm days, and a great shortage of sunshine. Yields of grain, fruit and root vegetables were low in most parts of the UK.

8 November to 23 December   *Severe gales; floods*
This exceptionally windy period produced five destructive gales; on 11–12 November peak gusts were 101mph at Stornoway and 86mph at Millport, on 26 November the Lizard reported 98mph, on the

29th St Mary's in the Isles of Scilly registered 105mph, on the 30th 107mph was recorded at Pembroke Dock, and on 21 December Kinloss reported a gust of 104mph. There was much damage on land, a large number of vessels foundered in home waters, and there was considerable loss of life at sea. On 27 November the tanker *World Concord* broke her back in the Irish Sea, another tanker sank with all hands off Cornwall, and the South Goodwin lightship sank with the loss of all but one of her crew. Widespread flooding affected the West Country, Wales, the Midlands and north-west England. Major storm surges along the east coast of England amounted to 2–2.5m on both 21 and 23 December, but were associated with neap tides, so a serious coastal flood was avoided. On 8 December a tornado caused a good deal of damage in west and north London, including severe damage to Gunnersbury Tube station.

17–18 December   *Heavy rain*
At Cruadhach, Loch Quoich, in the western highlands, 257mm fell in 22.5 hours.

1955

4–18 January   *Heavy snowfalls*
Snow fell frequently, widely and sometimes heavily during this period. Some 30–35cm lay in Somerset on the 14th, while level snow was reported 50cm at Glenrossal and 60cm deep at Elphin (both in Sutherland) on the 18th. Many towns and villages in northern Scotland, including Orkney and Shetland, were cut off by drifts up to 9m high, and the RAF and the Admiralty joined forces in 'Operation Snowdrop' to provide food, fuel and medical supplies.

9–28 February   *Heavy snowfalls*
'Operation Snowdrop' was revived when blizzards returned to northern Scotland after a much milder interlude. Heavy snow also fell in north Wales, the north Midlands and north-east England. Maximum snow depths were 92cm at Drummuir (Banffshire), 76cm at Bwlchgwyn (Denbighshire), 71cm at Glenlivet (Banffshire), 66cm at both Elphin (Sutherland) and Glenmore (Inverness-shire), 61cm at Braemar (Aberdeenshire) and 51cm at Buxton (Derbyshire). At Braemar the temperature fell to –25°C on the 23rd.

23 March *Severe gale*
Much damage on land in south-west England, and many losses at sea, including the sinking of the Norwegian vessel *Venus*, in Plymouth Sound. Highest gust was 94mph at St Mary's.

20–21 March *Heavy snowfalls*
Another blizzard in northern Scotland dropped 41cm at Reay Forest in Sutherland.

4–5 May *Duststorm*
Probably the most severe 'Fen Blow' of the century when south-westerly winds gusting to 65mph eroded topsoil from fields in Cambridgeshire, Suffolk and Norfolk, reducing visibility to 200m or less. Young crops were destroyed.

17–18 July *Flooding*
On the 17th serious flooding affected the Maidstone district after 107mm fell at Boxley and 106mm at Barming. On the 18th Dorchester and Weymouth suffered major floods following the highest authenticated one-day rainfall in the UK – 279mm at Martinstown, nearly all of which fell in ten hours. Nearby, 241mm fell at Friar Waddon and 229mm at Upwey, and a subsequent investigation suggested that as much as 325–350mm may have fallen locally. More than 100mm of rain fell over an area of 815 sq. km. At least fourteen people were killed by lightning during the month of July.

18–21 October *Flooding*
Widespread floods in Kent, Hampshire and Dorset. Poole collected 102mm in 24 hours on the 18th–19th, while on the 21st, 111mm fell at Ramsgate and 101mm at Broadstairs.

16–19 November *Smog*
Birmingham and the Black Country, Greater Manchester and Merseyside were all badly affected by a prolonged combination of fog and pollution which lasted over 72 hours.

20 December *Heavy snow*
A snowstorm disrupted transport and communications in northern England. Some 30–35cm fell in Yorkshire.

1956

10 January   *Heavy snow*
Snow fell widely, but was particularly heavy in Lincolnshire where
30–40cm fell.

23–25 January   *Heavy snow*
Another widespread snowfall, this time heaviest in north Wales:
Clawdd-Newydd had 46cm.

February   *Severe cold; heavy snow*
One of the coldest months of the century. Between 1,000 and 2,000
died from hypothermia as the temperature remained well below
zero between 31 January and 3 February, and from 18–25 February
inclusive. Snow fell frequently but heavy falls were localized. These
included 35–45cm in east and south Kent and in east Sussex;
25–35cm in the eastern half of Yorkshire, and 25cm in parts of
Cornwall – all in the third week. Drifts were up to 2m deep.

11 June   *Flooding*
Torrential rain caused serious flooding in West Yorkshire, especially
Keighley, Bingley and parts of Bradford and Halifax; 189mm fell at
Hewenden reservoir, above Bradford, of which 155mm fell in 110
minutes.

18–19 July   *Thunderstorms; flooding*
Severe flooding in south-west London followed falls of 109mm at
Harmondsworth and 98mm at Staines. Floods were also reported
from Blackpool, and from the Huntingdon/St Ives district.

29–30 July   *Severe gale; flooding*
The most severe and destructive high summer gale of the century.
Eleven people died on land, and many more at sea. Hundreds of
trees were uprooted, and some structural damage occurred. Highest
gusts included 93mph at the Lizard, 87mph at St Mawgan and
81mph at Dover. A severe flood also affected north-east Scotland,
including Lairg, Dingwall, Inverness and Nairn; 233mm fell in 48
hours at Arclach, Nairnshire.

6 August   *Hailstorm; flooding*
A prolonged thunder-and-hailstorm hit the Tunbridge Wells district; floodwater carried hailstones into the low-lying town centre, resulting in piles of ice 1 to 1.5 metres high.

12 December   *Severe gale*
Northern Scotland was swept by a damaging gale; a gust of 127mph was recorded at the exposed site at Costa Hill, Orkney, with 110mph also at Stornoway and 103mph at Lerwick.

25–26 December   *Heavy snow*
Deep snow disrupted traffic over the Christmas holiday; 20–25cm fell in Essex, the west and north Midlands and north Wales.

1957

31 January to 5 February   *Severe gales*
On the 31st western and central Scotland, north-west England and Northern Ireland were battered by a south-westerly gale gusting to 113mph at Tiree and 98mph at Eglinton (County Londonderry). On the 5th northern and western Scotland were worst hit with gusts to 109mph at Benbecula and 107mph at Tiree.

18–19 February   *Heavy snow*
Localized heavy falls disrupted traffic in the east and north Midlands and Yorkshire. Snow lay 38cm deep at Redmires in York-shire and 30cm in Leicester.

18–20 April   *Gales; sandstorm*
On the 20th a westerly gale gusted to 97mph at Stornoway; sand-storms moved vast tracts of dunes along the Nairn and Moray coasts.

8 June   *Flash flood; thunderstorm; heavy hail*
A tremendous downpour amounting in total to 203mm, of which 140mm fell in 150 minutes, hit Camelford in Cornwall. Wadebridge was also badly affected. There was much damage to roads, bridges and buildings, and hail accumulated to a depth of more than 0.5m.

5–12 August   *Thunderstorms; floods*
On the 5th, 105mm of rain fell in 90 minutes in Hereford, while that night 152mm was registered in eight hours at Rodsley in Derbyshire. Over 1,000 homes were flooded, as was a steelworks in Scunthorpe. On the 10th a fall of 136mm was recorded in just over two hours at Llansadwrn, Anglesey, and 98mm fell at Bangor. Flooding occurred widely in north Wales, and a railway embankment collapsed near Bangor. It was London's turn on the 12th when 99mm fell in a storm at Hampton.

17 September   *Flooding*
Floods hit north-west England after 119mm fell at Gosforth, Cumbria, and 97mm fell at Abbeystead, Lancashire.

26–29 October   *Flooding*
North Wales suffered widespread flooding. At Blaenau Ffestiniog, 323mm fell in four days.

3–4 November   *Severe gale*
Hundreds of new houses lost their roofs in Hatfield. West Raynham, Norfolk, had a gust of 104mph, and Dover one of 92mph.

3–5 December   *Smog*
A three-day smog episode in London; 90 people died in the Lewisham train crash.

1958

19–24 January   *Heavy snowfalls*
Widespread snow, heavy over large areas, caused major disruption to traffic. Helicopters were used to deliver supplies to villages in East Anglia. Notable level snow depths included 41cm in Ross and Cromarty, 38cm at Clacton and Walton-on-the-Naze, 59cm at Shoeburyness and 69cm at Aldergrove airport in County Antrim. Drifts of 4 to 5m were widely reported.

24–26 February   *Heavy snowfalls*
Heavy snow fell widely with severe drifting. Depths of 25–35cm

occurred widely, with 41cm at Bolton and Sheffield. Drifts to 5m were noted.

6–14 March   *Heavy snowfalls*
Northern England and Scotland were worst hit, with 36cm at Ampleforth and Redcar, and reports of 60cm in upland parts of Scotland.

3 June   *Flash flood*
Nearly 100mm of rain fell in west Cornwall leading to flash floods in Camelford, Wadebridge and Boscastle. Dozens of properties were affected.

1–3 July   *Severe flooding*
A torrential downpour over the southern Pennines (no gauge measurements, but estimated to be 100–125mm) led to dramatic flooding along the Sheaf and Don rivers. Hundreds of livestock were drowned. Thousands of homes, and also the main railway station, were flooded in Sheffield where it was considered to be the worst since the great flood of 1864.

6–9 September   *Thunderstorm; heavy hail; flooding; tornadoes*
A truly dramatic series of storms affected Sussex, Surrey, Kent, parts of Greater London and Essex. Highest rainfall totals, most of which fell in two hours, were 131mm at Knockholt and 102mm at Eynsford, both in Kent. Two tornadoes caused a good deal of damage in the Horsham/Gatwick district, and the UK's largest authenticated hailstone, 10cm across and weighing 141g, fell at Horsham.

1959

8–11 January   *Heavy snow*
Widespread falls caused much disruption to travel by road, rail and air. Norwich reported level snow 41cm deep, with 30–40cm in Sutherland, Caithness and North Yorkshire. Drifts of 6m isolated villages in northern Scotland. A further storm delivered 30cm to north-east Scotland between the 23rd and 25th.

February to October   *Drought*
This extended dry period, allied to a very warm and sunny summer, led to a severe agricultural drought. Most regions had restrictions for domestic water users as well. Averaged over England and Wales there was a 40 per cent shortfall, and locally in eastern England the deficit was close to 50 per cent. The rains returned in late October.

9–11 July   *Thunderstorms; large hail; flooding*
A zone stretching from Hampshire across the western and northern Home Counties to East Anglia experienced violent thunderstorms and torrential if short-lived downpours. Hailstones 5cm across fell at Wokingham. At Hindolveston, Norfolk, 88mm fell in four hours, of which 63.5mm fell in 20 minutes.

10 August   *Thunderstorms; flash flood*
Flash floods and landslips hit north-west Somerset, north Devon and central Cornwall after dramatic downpours. Nutscale reservoir, above Porlock (Somerset), recorded 130mm, mostly in five hours, and an estimated 200–250mm may have fallen in the town. In the Cornish storm 121mm fell at St Mawgan airfield, above Newquay. Severe local flooding was also reported from the Birmingham area, Hampshire, the Home Counties and East Anglia. More than 100mm of rain fell over an area of 450 sq. km.

1960

15–19 January   *Heavy snowfalls*
Southern Britain was badly hit on the 15th–16th with 25–35cm across a broad zone extending from Wiltshire and Hampshire to Bedfordshire and Northamptonshire. Scotland and northern England had some heavy falls on the 18th–19th with around 60cm in the Scottish Borders, Aberdeenshire, Ross and Cromarty, and Sutherland. The east coast main railway line was closed for a time. Further south a rapid thaw triggered flooding.

22–24 June   *Thunderstorms; flooding*
Floods were particularly bad in Oxfordshire and Berkshire, and two car factories – in Cowley and in Abingdon – had to be closed down.

At least two people were killed by lightning. Just outside Oxford 101mm of rain fell in five hours.

6 July   *Flooding*
Serious floods in central Lancashire and West Yorkshire; 129mm fell at Haredon Valley.

9–11 August   *Flooding*
Severe floods in Brighton and Worthing; 125mm of rain was recorded in 72 hours in Brighton.

24–25 August   *Flooding*
Major floods hit the north coast of Ulster, with 150mm at Coleraine and 122mm at Portrush on the 24th. The next day flooding occurred around Forres, Elgin, Nairn and Inverness.

7 October   *Flash flood*
A torrent of water tore through the centre of Horncastle, Lincolnshire, carrying cars, buses, household furniture and the stock of several shops. One man was drowned. At Horncastle, 184mm of rain was collected in just over five hours; 163mm fell at nearby Revesby.

27 September to 4 December   *Repeated severe flooding*
Prolonged heavy rain fell repeatedly over south and south-west England and south Wales resulting in several bouts of disastrous flooding, notably in south Devon including Exeter and Torbay. Taunton and Bridport were also very seriously affected, as were the Welsh Valleys in early December. The Taw, Torridge, Exe, Culm, Tone and Ogmore rivers were all in flood for several weeks. Notable rainfall events included 103mm at Trevone, Cornwall, on 27 September; 99mm at Colyton, Devon, on 30 September; 175mm at Teignmouth in the five days ending 2 October; 103mm at Luxborough, Somerset, on 8 October; and on 3 December Exford had 117mm, Simonsbath 107mm, Neuadd reservoir (Brecon) 154mm, Lluest Wen reservoir 153mm, Treherbert 151mm and Treorchy 149mm. In all, some 2,000 homes in the West Country were flooded, many damaged beyond repair, and 100mm or more fell over an area of approximately 500 sq. km.

1961–70

1961

27 January *Severe gale*
A severe south-westerly gale blasted across northern Scotland causing considerable damage; highest gusts included 109mph at Lerwick, 107mph at Stornoway and 103mph at Kirkwall.

26 February *Severe gale*
A widespread gale hit northern Britain, but destructive winds were confined to a fairly narrow belt across central Scotland. Tiree recorded 116mph and Lossiemouth 102mph.

27–31 May *Damaging frosts*
Several nights of air frost damaged young crops; Santon Downham (Norfolk) logged –6.7°C.

16–17 September *Severe gale*
A depression containing the remnants of Hurricane 'Debbie' tracked across Ireland and Scotland, causing widespread and severe structural damage, and a good deal of landscape damage. Highest gusts included 106mph at Ballykelly (Co. Londonderry) and 103mph at Tiree.

28 December to 1 January 1962 *Heavy snow*
Two heavy falls of snow affected large parts of the UK, but worst affected were the northern Home Counties where 30cm or more fell widely, and as much as 41cm over the Chilterns.

1962

11 January *Severe gale*
Damage was worst in south-west England; highest gust was 104mph at Hartland Point.

21 January *Severe gale*
Northern Scotland was badly hit with a gust of 102mph recorded at Stornoway.

12–18 February   *Severe gales*
Northern Scotland was again badly affected on the 12th with a gust
to 113mph at Stornoway. On the 16th–17th gales affected all areas,
but Sheffield was particularly badly hit with almost 70 per cent of
the city's housing stock damaged; almost 100 homes had to be
demolished. A gust of 109mph was reported from Kirkwall, and one
of 97mph in Sheffield itself.

2 March   *Heavy snow*
North-east England and the Scottish Borders were hardest hit; 38cm
of snow lay at Haydon Bridge.

16 May   *Severe gale; flooding*
Northern Scotland suffered both gale and flood damage. Benbecula
(Western Isles) recorded a gust of 100mph, while 136mm of rain fell
at Gobernuisgach Lodge in Caithness.

29 May–8 June   *Late spring frosts*
Air frost affected parts of England on eleven consecutive nights,
causing extensive horticultural and agricultural losses. Santon
Downham (Norfolk) recorded –5.6°C on both 1 and 3 June.

6 August   *Flooding*
A bank holiday deluge led to flooding from Dorset northwards to
the Black Country. Highest rainfall totals were 102mm at Chipping
Campden and 100mm at Blockley, both Gloucestershire.

1–7 December   *Smog*
Although in some respects this was the foggier period of the two, the
death toll was much lower than during the 1952 smog, an indication
of the effect that the Clean Air Act of 1956 was already having. Nevertheless, almost 1,000 deaths were attributable to the pollution in
London alone. In parts of the capital the fog persisted without a
break for 96 hours.

15–16 December   *Severe gale*
All districts were swept by damaging winds during these two days;
highest gusts were 98mph at Tiree, 95mph at Rannoch, and, further
south, 90mph at Bidston observatory.

1963

22 December 1962 to 4 March 1963   *Severe winter; heavy snowfalls*
This was the coldest winter of the twentieth century, and severe weather persisted without a break for over ten weeks. The temperature remained below zero for long periods, especially in central, eastern and southern parts of England, and the number of deaths from hypothermia was estimated as being between 3,000 and 5,000. Agriculture was seriously affected, the number of registered unemployed trebled, and there was considerable frost damage to roads and railways. There were also frequent and long-lasting failures to electricity and gas supplies, and coal for domestic fires was in short supply. Snow fell less frequently and less heavily than in 1947, but there were a number a disruptive falls, and much of England was snow-covered from 26 December to 2 March. Major falls, each contributing 25cm over a wide area, occurred on 26–27 December, 29–30 December, 3–4 January, 16 January, 19–20 January, 5–7 February and 14 February. Noteworthy snow-depths included 35–45cm over much of Kent, Surrey, Sussex and Hampshire on 28 December, and large parts of the West Country on 30 December; 51cm at East Grinstead on 1 January; 60–65cm at Princetown on Dartmoor, with drifts to 8m, on 20 January; 165cm at Tredegar on 8 February; and 80–85cm in Redesdale, Northumberland, on 15 February. Freezing rain also occurred on occasion; probably the most widespread such event was across southern England on 3 January. Temporary thaws resulted in serious flooding in the West Country on 30–31 December and 7–10 February, but the severe weather this time ended with a slow thaw under sunny skies, and flooding in March was confined to northern England during the second week, from a combination of the thawing snow-cover in upland areas and heavy rain.

6–12 June   *Thunderstorms; flooding*
Thunderstorms occurred daily during this period leading to serious local flooding. On the 7th, 94mm fell at Mill Hill, north London, most of it in less than two hours. On the 11th, 93mm fell at Gortnamoyagh Forest in County Londonderry.

1964

13–14 January   *Heavy snowfall*
Snow fell widely across England and Wales, up to 30cm deep around Brighton and Worthing.

15–16 March   *Heavy snowfall*
Extensive snow, deepest in the Midlands and Yorkshire where 25–30cm fell.

28–31 May   *Thunderstorms; flooding*
Widespread thundery activity, with some intense storms over the central Pennines on the 30th, leading to flooding in Yorkshire and Lancashire. Brushes Clough, near Oldham, collected 123mm.

18–21 July   *Thunderstorm; hailstorm*
On the 18th, 56mm of rain fell in fifteen minutes in Bolton, while the moors around Pateley Bridge were 15cm deep in hailstones. On the 21st, 105mm fell at Botolph Claydon, Buckinghamshire.

9 October   *Severe gale*
A ferocious gale in the Channel Islands destroyed several hectares of glasshouses and caused extensive damage to buildings and communications. Jersey airport logged a gust of 108mph.

14 October   *Flooding*
East Kent suffered severe flooding; 94mm of rain fell at Martin Mill, north of Dover.

7–13 December   *Severe flooding*
Flooding in mid- and north Wales was described as the worst of the century. On the 12th alone, 173mm was recorded at Ceinws and 188mm at Blaenau Ffestiniog. Weekly totals exceeded 500mm at several sites.

1965

13–17 January *Severe gales*
Damaging gales affected southern Scotland, northern England and Northern Ireland on the 13th, with peak gusts of 93mph at Ronaldsway, Isle of Man, and 92mph at Carlisle. On the 17th Wales, the West Country and the Channel coast were badly hit, with gusts of 101mph at Portland Bill, 97mph at Valley, Anglesey, and 95mph at the Lizard.

22 January *Heavy snow*
Heavy snow fell across most of England and Wales north of London and Bristol; snow lay 20–30cm deep in parts of Derbyshire and Staffordshire and 46cm deep at Tredegar, south Wales.

1–5 March *Heavy snow*
Snow fell heavily in north-east England on the 1st, with depths typically 20–30cm, but on the 4th widespread snow affected most of England and Wales, accumulating to a depth of 20–30cm in parts of Wessex and the Midlands, and 35cm in mid-Wales (37cm at Evancoed, Radnorshire), with drifts of 2–3m. Snow returned to northern England on the 21st when Buxton reported 36cm.

11–24 July *Thunderstorms; flooding; tornado*
This very thundery period produced floods in Devon on the 12th, in Yorkshire and Lancashire on the 13th, in Cornwall on the 14th, in London and the northern Home Counties on the 20th, in Lancashire again on the 23rd, and in south-west Scotland on the 24th. A small tornado wreaked havoc in the Royal Horticultural Society's gardens at Wisley on the 21st. The most notable rainfall total was 139mm at Trevanson, near Wadebridge, on the 14th, which fell in less than four hours.

1 November *Severe gale*
Three of the eight cooling towers at Ferrybridge, near Doncaster, collapsed in a westerly gale; the remaining five were all damaged. This disaster was more a consequence of bad planning than of unusually high winds, although a gust of 93mph was noted at Carrigans, County Tyrone, during the same gale.

13 November to 2 December   *Heavy snowfalls*
Snow fell widely and frequently during this period, with major accumulations in northern England and southern Scotland during the last few days of November. Ushaw, near Durham, reported 56cm of level snow at this time. Thaw floods followed in the first half of December.

17–18 December   *Severe flooding*
Exceptional amounts of rain fell over a 48-hour period over the hills of south Wales, resulting in catastrophic floods in the Welsh Valleys. Exmoor was also badly hit. Highest two-day totals were 285mm at Lluest Wen reservoir, in the upper Rhondda Valley, and 271mm at nearby Lluest Wen filters. There were four other totals above 200mm, and one of 177mm at Simonsbath on Exmoor.

1966

2 January   *Severe gale*
Widespread gales, worst in the West Country; gusts to 102mph at Portland Bill.

11–20 January   *Heavy snow; freezing rain*
Heavy snow fell on the 11th in Wessex and south Wales, with a level depth of 38cm at Tredegar, while a general snowfall on the 15th produced depths of 25–30cm in Kent. Freezing rain fell widely on the 20th across southern England, including London where hospitals reported that they had treated hundreds of casualties for fractures and bruising.

14–21 February   *Heavy snow; thaw floods*
A disruptive snowfall in north-east England with 30cm at Ushaw, near Durham. A rapid thaw on the 20th–21st accompanied by heavy rain led to widespread floods in northern England.

27 March   *Severe gale*
Damaging winds affected much of the UK; highest gust was 98mph at Bidston observatory in Cheshire.

1 April   *Heavy snow*
Some 30–35cm of level snow fell over large parts of north Wales and north-west England, severely disrupting transport. Two weeks later, a notably late snowfall on the 14th–15th dropped some 15cm over southern England.

August   *Various flooding events*
Flooding in Northumberland, the Borders and the Lothians followed a prolonged downpour on the 3rd–4th during which 102mm fell at Whittinghame, East Lothian, in 48 hours. On the 13th Northumberland was again in the firing line, this time accompanied by Dumfriesshire, Galloway and Lanarkshire; 107mm fell at Sweetshaw Burn (Lanarkshire) and 102mm at Eliock (Galloway). Severe thunderstorms triggered flooding in parts of Gloucestershire and Staffordshire on the 20th, while a broad zone extending from Hampshire to Birmingham was affected on the 29th – August bank holiday – with a fall of 105mm at Wroxhall (Warwickshire).

16–31 October   *Flooding; tornado; Aberfan disaster*
Repeated bouts of heavy rain affected many parts of the country including south Wales where on the 21st a saturated spoil-heap collapsed and engulfed several buildings, including a school, in Aberfan, killing 144 people of whom 116 were children; although rain triggered the collapse, the chief culprit was inadequate management of the artificial hill. A tornado caused damage at Headington, Oxford, on the 16th, while five days of heavy rain led to flooding in east Kent at the end of the month with 125–140mm of rain falling in 120 hours.

17 December   *Severe flooding*
Prolonged heavy rain combined with a rapid thaw led to widespread floods in the south-west Highlands; at Dalness in Argyllshire 199mm fell on the 17th alone, and 270mm in 48 hours.

1967

13–14 July   *Thunderstorms; large hail; flooding*
Serious flooding affected parts of Wiltshire, Merseyside and the

Wirral, and Northern Ireland. A violent hailstorm damaged glass-houses, vehicles and roofs in and around Chippenham (Wiltshire).

22–23 July   *Flooding*
There was extensive flooding in the Home Counties and also south Devon following a heavy overnight downpour. West Byfleet (Surrey) collected 100mm, and Metherall (Devon) 97mm.

8 August   *Severe flooding*
A damaging flood hit parts of east and central Lancashire; 117mm of rain fell in 80 minutes at Dunsop Valley in the Forest of Bowland. There were several lightning deaths in other parts of England on the same date.

16 October   *Severe flooding*
Widespread and long-lasting floods hit much of mid- and south Wales and the western Midlands following an exceptional down-pour over the Welsh hills. Beacons reservoir (Breconshire) collected 139mm and nearby Storey Arms Field 134mm.

8–9 December   *Heavy snow*
Two polar lows brought two heavy snowfalls within 36 hours to Northern Ireland, Wales, the Midlands and parts of southern England, seriously disrupting transport. Some 45–60cm of snow lay in mid- and north Wales, 25–30cm in Northern Ireland and the Midlands, and 28cm even at Brighton where the coast road was littered with abandoned vehicles.

1968

8–9 January   *Severe snowstorm*
The first major event of this year of disasters caught forecasters by surprise and deposited 30cm or more of level snow over Wales and a large part of England, with 40–50cm inland in mid- and north Wales, and drifts of 3m or more. Three snowploughs became stuck in drifts in west Berkshire. A rapid thaw on the 13th–14th led to a good deal of flooding.

14–15 January *'The Glasgow hurricane'*
An intense depression tracked across northern Scotland; there were widespread gales which were particularly destructive in the Central Belt of Scotland where extensive tracts of forest were flattened, power and telephone lines were brought down, and structural damage occurred widely. The Glasgow conurbation was severely hit and over 1,000 homes were badly damaged, 50 of which had to be demolished. Twenty people died, a further 40 were injured, and over 600 had to be given temporary accommodation. Peak gusts included 117mph at Tiree, 106mph at Leuchars, 103mph at Prestwick, Edinburgh airport, and 102mph at Glasgow airport.

5 February *Heavy snow*
Another unpredicted snowfall hit much of the western Midlands, Cheshire and Lancashire, with many places reporting over 25cm of undrifted snow; Keele University, Buxton and Onecote (Staffordshire) had 35–40cm. The urban areas of Birmingham and the Black Country, the Potteries, Merseyside, Greater Manchester, Preston and Blackpool were particularly badly affected with complete dislocation of transport, while power and telephone lines were brought down by the soft, clinging nature of the snow.

17–18 March *Duststorm*
A severe Fen Blow affected Cambridgeshire, Suffolk and Norfolk, with peak gusts of 65mph.

23–27 March *Widespread floods*
Widespread floods hit many parts of western Scotland following record-breaking downpours on the 26th–27th. Kinlochewe (Wester Ross) collected 253mm in 36 hours and Broadford (Skye) 248mm in the same period. Approximately 12,500 sq. km received over 100mm of rain. Three days earlier, north Wales experienced a major downpour with 152mm in 24 hours at Cowlyd in the upper Conway catchment, triggering flooding there.

1–2 July *Thunderstorms; hailstorms; flooding*
Violent thunderstorms swept the West Country and south Wales late on the 1st; hailstones 6–8cm across fell in Slapton (south Devon) and at Cardiff airport. Severe flooding affected towns and villages throughout

Devon. Early on the 2nd the storms moved to northern England, the Isle of Man, where 148mm of rain fell at West Baldwin reservoir, and south-west Scotland, where Lagafater Lodge in Galloway collected 99mm. Bulldozers were used to clear hail 20cm deep from streets in Yorkshire. At least five people were killed by lightning.

9–10 July   *Severe flooding*
Many places the length and breadth of England and Wales suffered flooding to some degree; these floods were disastrous and long-lived in Bristol, Bath, Gloucester and Peterborough. In Somerset the Cheddar Gorge caves were flooded for the first time in living memory. The heaviest fall was 173mm, most of which came down in six hours, at Chew Stoke in Somerset, but more than 100mm fell at several sites in Somerset, Gloucestershire, Worcestershire, Warwickshire, Northamptonshire, Cambridgeshire and Lincolnshire. In all, 2250 sq. km collected over 100mm of rain.

14 July   *Flash flood*
A total of 100mm of rain fell at Llanfyllin in Montgomeryshire and torrents of water rushed through the village. Downstream flooding occurred along the Afon Cain and the River Severn in Shropshire.

13–16 September   *Severe flooding*
After two moderately wet days, it rained almost without a break throughout the 15th, the day's total amounting to 201mm at Tilbury and 200mm at Stifford. Some 6,250 sq. km of land, chiefly in a broad zone stretching from Hampshire and north-west Sussex across Surrey, north Kent and south Essex, received over 100mm of rain. All routes out of London to the south-west, south, south-east and east, became impassable, and thousands of homes were flooded, especially in south-west London and north Surrey. It took over a week for the floodwaters to subside.

20 September   *Flooding*
Floods affected many places in mid- and east Lancashire and in west and south Yorkshire, following a major downpour over the central Pennines. Some 117mm of rain fell at Great Heys and 115mm at Yateholme.

31 October to 1 November   *Severe flooding*
Heavy rain fell throughout Northern Ireland during these two days, with more than 100mm registered over more than 1,000 sq. km. At Tollymore Park, on the northern flank of the Mourne mountains in County Down, 240mm fell in the 48 hours, of which 159mm fell on the 31st alone. Flooding was extensive, especially in the Upper Bann catchment, and in Belfast.

24–31 December   *Heavy snow; freezing rain*
Rain unexpectedly turned to snow on Christmas Eve across much of Wales and the Midlands and parts of southern England, and the following morning snow lay 15–25cm deep in parts of the Midlands and Wales. In west and south Wales freezing rain fell. Further heavy falls affected chiefly east coast counties of Scotland and England during succeeding days with undrifted depths of 48cm at East Dereham, Norfolk, and 40cm at High Mowthorpe, East Yorkshire. Whitby was cut off for almost a week by 3m drifts.

1969

17 January   *Severe gale*
The Lyme Regis inshore lifeboat capsized with the loss of one crew member. St Mary's, Scilly, reported a maximum gust of 91mph.

7–8 February   *Severe gale; heavy snow*
Dry, powdery snow fell over much of the UK, drifting readily in the strong wind. Level snow 30cm deep was reported from east Kent with 6m drifts; 20–25cm fell widely elsewhere. An exceptional low-level gust of 136mph was recorded at Kirkwall, and with the temperature standing at –4°C, extreme wind-chill pertained. On the 9th 25–30cm of snow fell in Ulster.

19–22 February   *Heavy snow; gale; coastal flooding; thaw floods*
Some 20–25cm of snow fell widely across England and Wales, with 35cm at Tredegar; drifts up to 5m were reported. Several people were injured in avalanches in the Cairngorms. An easterly gale, gusting to 82mph at the Lizard, caused extensive coastal flooding in south Devon, especially around Teignmouth, Dawlish and

Paignton. Roads and railways were impassable. A rapid thaw on the 20th–22nd led to serious river flooding in the West Country.

16–19 March *Heavy snow; freezing rain*
The 384m-high television transmitter at Emley Moor collapsed as a result of a combination of heavy icing and strong winds. Freezing rain fell widely in the Midlands and northern England, seriously affecting road and pedestrian traffic, while accumulations of glaze and rime amounted to 15cm in heavily populated parts of West and South Yorkshire. Elsewhere in northern England 20–30cm of snow fell. The Longhope lifeboat foundered off Orkney in a gale on the 18th with the loss of all eight crew.

28–29 July *Heavy rain; flooding*
Widespread but short-lived floods affected much of Cornwall, Devon, Dorset and Somerset, and some districts elsewhere, following record-breaking rainfall in the West Country over a period of about fifteen hours. More than 100mm of rain fell over an area of 5,250 sq. km, with maximum falls of 145mm at Ellbridge, opposite Plymouth, 144mm at North Hessary Tor and 143mm at Southcott Cross, both on Dartmoor, and 141mm at Broadhembury, near Honiton.

2 August *Thunderstorms; flooding*
Widespread thunderstorms over England and Wales led to flooding, particularly in west and south London, in east Berkshire, in Norwich and in south Warwickshire. Southam (Warwickshire) logged 127mm and Kempton (Surrey) 102mm.

28–29 September *Severe gale; storm surge*
A violent west to north-westerly gale caused widespread damage in northern Scotland where Dounreay recorded a gust of 111mph, Wick one of 102mph, and Kirkwall one of 101mph. An associated storm surge of 2–2.5m affected the east coast of England, with flooding reported to be the worst since the 1953 disaster.

21 November *Severe flooding*
Some 60–100mm fell widely in Northern Ireland, with 138mm at Ballybraddin Forest in County Antrim. Omagh was completely cut off by flood waters for two days.

1970

18–21 January   *Avalanche; rough seas*
On the 18th three died in an avalanche on Ben Nevis, and on the 21st the Fraserburgh lifeboat capsized in rough seas with the loss of five crew members.

12 February   *Snowstorm*
Widespread snow, drifting badly in the strong east wind, seriously disrupted transport. Some 15–25cm fell widely over central and southern parts of England and Wales, with 30cm locally in upland parts of south Wales. On the 7th an avalanche in Glencoe claimed one life.

4 March   *Heavy snow*
An unexpected snowstorm affected a broad swathe extending from Northern Ireland, across north-west England, much of Wales, the Midlands, East Anglia and the south-east. Undrifted snow was 48cm deep at Sywell aerodrome, near Northampton, 36cm deep at Thurleigh, north of Bedford, and 30cm deep at Farthing Common, Kent. Miners in the Kent coalfield were trapped underground for several hours following a power failure.

22 April   *Flooding*
Widespread heavy rain led to floods in northern England, Wales and the Midlands. The rain was heaviest in upland areas, and in Cumbria 182mm fell at Seathwaite and 163mm at Honister Pass.

7–11 and 26–27 June   *Thunderstorms; hailstorms; flooding*
Severe thunderstorms broke out widely between the 7th and 11th with some very intense short-period downpours. On the 7th, 93mm fell in two hours at Lossiemouth in Morayshire; on the 10th, 111mm fell at Miserden in Gloucestershire and 127mm at Drimsyniebeg, above Loch Goil, in Argyllshire; and on the 11th 93mm fell at Hampton Park, Worcestershire. Violent hailstorms ruined thousands of acres of crops and hundreds of glasshouses in Leicestershire and Cambridgeshire on the 27th, and at Wisbech 100mm of rain was recorded in 140 minutes, of which 51mm came down in just twelve minutes. At least five people were killed by lightning.

24 July   *Flooding*
Large parts of Aberdeen and the surrounding district had to cope with floodwaters over a metre deep following a downpour which dropped 115mm at Netherley, 109mm at Mannofield and 97mm at Craigiebuckler.

15–19 August   *Widespread flooding*
On the 15th–16th both Northern Ireland and north-east Scotland were badly hit. Large parts of the Belfast urban area suffered severe flooding when 118mm of rain fell at Kilroot and 111mm at Loch Mourne on the 15th alone. In Scotland, the Rivers Spey, Lossie and Findhorn all overtopped their banks leading to destruction of arable crops, drowned livestock and serious flooding in the towns of Elgin and Forres. The Aberdeen–Inverness railway line was washed out in several places. At Lochindorb Lodge, Morayshire, 152mm fell in 48 hours. On the 19th the focus of attention turned to Birmingham and Somerset, both of which suffered badly; 97mm fell at King Edward's School, Edgbaston, and 98mm at Timberscombe, Somerset.

25–28 December   *Heavy snow*
Snow fell widely over the Christmas holiday, seriously disrupting transport. Some 20–25cm accumulated in east and north Kent, and in upland parts of north-east England.

1971–80

1971

19 March   *Flooding*
A good deal of flooding occurred in Northumberland, the Scottish Borders and the Lothians – including Edinburgh – following a prolonged downpour which deposited 113mm at West Hopes (Northumberland), 111mm at Whiteadder reservoir (East Lothian) and 98mm at Penicuik (Midlothian).

24 April   *Flooding*
Heavy rain fell widely on this date causing a good deal of flooding; 130mm fell at Craggs Lane Farm, near Richmond, North Yorkshire.

3 July   *Thunderstorm; flooding*
Floods affected parts of Denbighshire, Flintshire and Cheshire, including the Chester and Wrexham districts, following falls of 97mm at Cadole, 95mm at Mold, and 81mm at Ruthin, mostly in less than three hours.

26 September   *Tornadoes*
Several tornadoes were reported; one caused extensive structural damage in Rotherham.

18–21 October   *Flooding*
Severe flooding affected parts of West Yorkshire, notably in and around Huddersfield, on the 18th; 103mm fell at Redbrook reservoir above Huddersfield. Prolonged heavy rain on the 21st in west and north-west Scotland produced falls of 158mm at Cassley power station (Sutherland) and 162mm at Dalness (Argyllshire).

30 November   *Thick fog*
Patches of thick, drifting fog, allied to incompetent driving, resulted in a multi-vehicle pile-up on the M1 near Luton. Seven were killed and 40 injured.

1972

18–19 July   *Thunderstorms; flooding*
Violent storms, some accompanied by heavy hail, affected in particular Devon and Kent. On Romney Marsh sheep were killed by hailstones, in the Exeter area houses were damaged by lightning and at Exeter airport 90mm of rain fell in 130 minutes. Flooding in the city was severe, if short-lived.

31 July–1 August   *Thunderstorms; flooding*
Large parts of Norwich were under water early on the 1st following overnight storms. At nearby Old Costessey 138mm of rain fell in less than four hours. A boy was killed by lightning.

5 and 13 December   *Severe gales*
Scotland was swept by gales; a gust of 101mph was recorded at Hunterston (Ayrshire) on the 5th, and one of 94mph at Lossiemouth (Morayshire) on the 13th.

1973

19 January   *Flooding*
Serious floods in Northern Ireland, especially in County Down, followed a prolonged downpour in which 115mm of rain fell at Castlewellan Forest.

21–22 January   *Heavy snow*
Snow fell widely from the Midlands northwards, with 35cm at Ashintully Castle (Perthshire).

12–15 February   *Heavy snow*
Heavy snow fell in northern England, Northern Ireland and Scotland, with 41cm at Fersit (Inverness-shire).

2 April   *Severe gale*
The physical impact of this gale, with structural damage and thousands of uprooted trees especially in the Midlands, East Anglia and parts of the Home Counties, seemed out of proportion to the recorded wind speeds. Highest gust was 76mph at West Raynham (Norfolk).

6 July   *Thunderstorm; flooding*
A violent but localized storm caused serious, if short-lived, flooding in south-west London and north-west Surrey. At Chessington, 117mm of rain fell in less than five hours.

15–16 July   *Severe flooding*
Widespread and disastrous flooding affected large parts of west and south Yorkshire, north Derbyshire, east Cheshire and Greater Manchester. Sheffield was particularly badly hit. There was also flooding in other parts of Yorkshire, Lincolnshire and Nottinghamshire. On the 15th alone, a total of 26 rainfall stations recorded over 100mm of

rain, with 137mm at Rivelin and 134mm at Derwent reservoir, both above Sheffield, and 119mm in Sheffield itself. Over the two-day period, during which the bulk of the rain fell in 42–44 hours, highest rainfall totals were 170mm at Derwent reservoir, 168mm at Howden reservoir, 165mm at Rivelin and 135mm in Sheffield, and the area affected by 100mm or more was 2,240 sq. km.

20 September   *Severe flooding*
A phenomenal deluge hit a large area covering north-eastern and central Kent, leading to destructive floods in Thanet, Sandwich, Deal, Canterbury and Ashford. In just over twelve hours, 191mm of rain was recorded at West Stourmouth, 172mm at Margate, 167mm at Ashford, 161mm at Manston and 151mm at Minster. In all, 32 rainfall stations recorded over 100mm.

12 October   *Flooding*
Severe flooding in west Cornwall followed a downpour which dropped 90–100mm of rain on five sites in and around Penzance.

19 December   *Heavy snow*
Snowstorms in the Scottish Highlands disrupted transport; 38cm fell at Glenmore Lodge.

1974

4 January to 6 February   *Frequent severe gales; flooding*
This was arguably the longest period of rough weather since the early 1930s and it was not again equalled until 1990. There were several destructive gales, and damage to buildings and landscape, transport networks, and overhead power and telephone lines was extensive. Peak gusts included 102mph at Port Talbot on the 8th; 123mph at Kilkeel (County Down) and 104mph at Lyneham (Wiltshire) on the 11th–12th; 100mph at Castle Archdale Forest (County Fermanagh) on the 28th; and 89mph at St Mary's (Isles of Scilly) on 2 February. Lightning, hail and tornado damage were also widely reported overnight 10th–11th, and flooding occurred widely, especially in western and central Scotland after 238mm of rain fell in 24 hours on the 17th at Sloy Main Adit (Dunbartonshire); in all, eight sites registered over 100mm that day.

16 June *Thunderstorms; flooding*
Widespread thunderstorms affected the Midlands and south; serious flooding was reported from the Bristol and Oxford areas; 96mm of rain was registered at Ambrosden (Oxfordshire).

7 August *Flooding*
Thunderstorms broke out widely across south-east England; 115mm fell at Fyfield (Essex).

2 September *Gale; heavy seas*
Morning Cloud III, the yacht belonging to former Prime Minister Edward Heath, foundered in heavy seas off Brighton with the loss of two crew members. Plymouth had a gust of 71mph.

Autumn *Flooding*
Flooding occurred in various parts of the country in each of the three autumn months, but it was particularly widespread during the third week of November in southern districts; at the time this was the wettest autumn since 1960.

11 and 29 December *Severe gales*
Structural damage was reported from the London area on the 11th when Kew Observatory recorded its highest gust to date of 89mph; aircraft were damaged at Heathrow. On the 29th Eskdalemuir reported a maximum gust of 95mph.

1975

2–28 January *Frequent gales*
Another very rough January brought frequent gales, some severe. There was a good deal of damage and disruption on land, and several vessels foundered in British waters. The death toll on land and at sea was at least 22. The peak gust was 90mph at the Lizard on the 27th.

27 March to 10 April *Heavy snowfalls*
This lengthy cold period brought frequent snow which disrupted transport, especially during the Easter holiday (28–31 March).

Birmingham was gridlocked on the 27th when 15–20cm of snow fell. There were similar falls in eastern England the following week, and one of 47cm at Glenmore Lodge, Inverness-shire, on the 8th.

29 May to 4 June, also 28–29 June   *Damaging frosts; some snow*
Sharp night frosts damaged fruit blossom and spring vegetables; a minimum of –5.1°C was registered at Leadhills, Lanarkshire on the 31st, and –3 to –4°C as far south as Buckinghamshire. Snow fell widely on 2 June. Further widespread frosts on 28 and 29 June damaged potato and soft-fruit crops.

17 July   *Thunderstorm; flooding*
A localized storm in Norfolk deposited 121mm of rain in two hours at Aylsham.

31 July   *Thunderstorm; flooding*
A violent thunderstorm in Hampshire dropped 96mm at Medstead and 85mm at Alton.

14 August   *Thunderstorm; heavy hail; flash flood*
Known as the Hampstead storm, this 'supercell' thunderstorm was practically stationary over a limited part of north London centred on Hampstead which was deluged with 171mm of rain, nearly all in 150 minutes. Nearby Golders Hill Park had 131mm, and four other gauges in the district caught 100mm or more. Torrents of water and hail poured down major roads and into thousands of homes, and the London Underground was seriously affected. One man drowned.

29 August   *Flooding*
Heavy rain fell widely, with 108mm at Sweethope Lughs in Northumberland, and 89mm at Busson Moor in Cornwall. Roads and homes were flooded in both districts.

13–14 September   *Severe flood*
A prolonged and widespread downpour affected the whole of southern England leading to extensive flooding. Dundry Hill, near Bristol, collected 100mm, Stanford-le-Hope in Essex 97mm, and Margate 91mm. Five deaths were associated with the bad weather.

10–16 and 29–30 November   *Widespread fog*
Fog formed widely during these two periods and was dense in places, leading to much disruption. Ten people died when two light aircraft crashed, attempting to land in poor visibility, one at Birmingham airport, the other at Elstree. Racing driver Graham Hill was one of the casualties.

Major drought
The most intense drought of the twentieth century stretched over 16 months between May 1975 and August 1976, and was exacerbated by long periods of hot weather with low humidity during the two summers. It culminated during the summer of 1976 in water restrictions for millions, poor crop yields, frequent heathland fires and a transformation of England's green and pleasant land into a bleached and burnt semi-desert. Mr Dennis Howell, a junior minister in the Labour government, was given special responsibility for co-ordinating Whitehall's responses to the crisis, and was popularly known as the 'Minister for Drought'. Several rainfall stations in Sussex recorded less than 20mm of rain from June to August 1976 inclusive. Averaged over England and Wales, the 16-month rainfall total was 763mm, just 63 per cent of the long-term mean of 1,204mm.

1976

2 January   *Severe gale; storm surge*
The twentieth century's most violent and destructive windstorm to hit England and Wales up to 1987 delivered sustained winds of 55–60mph to inland regions, and peak gusts included 105mph at Wittering (near Peterborough), 98mph at Coltishall (near Norwich), 97mph at Wattisham (near Ipswich) and Cilfynydd (in the Welsh Valleys) and 95mph at Cardington (near Bedford) and at Sheffield. There was widespread structural and landscape damage, especially in the Midlands and East Anglia, power and telephone lines failed, and transport on land, by sea and by air was seriously disrupted. Storm surges affected both east and west coasts, peaking at 2.5m at Southend, and there was some local coastal flooding, notably in Lincolnshire and Norfolk. The UK death toll was 24 (and in Europe, 60).

20–21 January *Severe gales*
Widespread gales, worst in Scotland, killed five people. Tiree recorded a gust of 95mph.

22 June to 16 July *Heatwave*
Starting on the 23rd there were fifteen days with a temperature of 32°C or more somewhere in England, and on five of those days the mercury passed 35°C. Many people died from heatstroke, and hundreds of others were hospitalized. This was also the first major ozone pollution episode in England.

30 August *Flooding*
The long drought broke with a downpour which caused serious if short-lived flooding, notably in parts of the Midlands and East Anglia. 106mm of rain fell at Thorney, near Peterborough. A USAF transport plane crashed near Thorney in a severe thunderstorm with the loss of seventeen lives.

11–28 September *Severe flooding*
Severe flooding on the 11th affected principally North Yorkshire and County Durham where 145mm fell at Kildale, 141mm at Easby, 140mm at Danby Lodge and 121mm at Sedgfield. Floodwaters nearly 2m deep flowed through the village of Stokesley. North-west Wales was also badly hit with 127mm at Capel Curig. On the 24th a flash flood caused structural damage to 80 buildings in Polperro (Cornwall) when 89mm of rain fell in 90 minutes, and the next day Somerset, Wiltshire and Gloucestershire were badly hit with 111mm at Stroud and 110mm at Bathford. On the 28th, deep floodwaters hit central Glasgow, gridlocking the city's roads, when 84mm of rain came down in less than four hours.

14 October *Severe gale*
A severe gale in the English Channel caused many difficulties for shipping. The Lizard recorded a gust of 83mph.

17–21 December *Heavy snow*
Snow fell widely across the UK, accumulating to a depth of 30–40cm in parts of Scotland. A 20cm fall in six hours caused traffic problems in Aberdeen.

1977

12–14 January   *Heavy snow*
A broad belt of heavy snow moved northwards across the country; accumulations included 25–30cm in east Kent and 40cm locally in the north Midlands and Yorkshire. A thaw followed. Seven deaths were related to the severe weather.

20–26 February   *Heavy snow; widespread flooding*
Although there were no individual daily falls of note, the monthly rainfall for February was between four and five times the normal in Derbyshire and Nottinghamshire, and flooding on the river Trent and its tributaries was severe and widespread; there were a further seven deaths. Nottingham was particularly badly hit. Heavy snow fell across upland parts of northern England with 20–30cm widely, and 61cm at Thornton Moor, West Yorkshire.

16–17 August   *Severe flooding*
West and north London and adjacent parts of Buckinghamshire and Hertfordshire were badly hit following falls of 115mm at Chalfont St Peter, 113mm at Maple Lodge and 113mm at Ruislip; hundreds of properties were also flooded in the Greenwich area. A second area of heavy rain led to widespread flooding also in Wiltshire, Gloucestershire and Somerset.

27 August   *Severe flooding*
South-east London and north-west Kent suffered seriously flooding after 97mm was recorded at Sidcup in less than five hours.

5 October   *Severe flooding*
Many homes were flooded in south Cornwall following falls of 99mm at Goonhilly Down, 94mm in Falmouth and 94mm in Truro.

8 October   *Tornado*
A tornado damaged over 60 houses in Grantham. Several minor injuries were reported.

30–31 October   *Severe flooding*
Cumbria, Galloway, Dumfriesshire and the Upper Tweed catchment

experienced serious flooding following a prolonged downpour which dropped 182mm in 24 hours at Thirlmere and 181mm at Braithwaite (both Cumbria), and 122mm at Ettrick School (Selkirkshire).

11–12 November   *Severe gale; coastal flooding*
A severe north-westerly gale produced a storm surge along the coasts of Lancashire and north Wales which breached sea defences, especially in the Fylde peninsula where 25–30 sq. km of land were inundated in the Fleetwood-Thornton-Cleveleys district. Liverpool and Morecambe both suffered considerable gale damage. The death toll was five. Fleetwood recorded a gust of 83mph.

27–30 November   *Smog*
Arguably the UK's last killer smog, a dense and heavily polluted fog descended on Glasgow and Clydeside for almost 100 hours. The death toll was between 200 and 300.

23–24 December   *Severe gale*
Southern England, south Wales, the Midlands and East Anglia experienced the strongest winds during this damaging gale. Peak gusts were 93mph at Cilfynydd (Glamorgan), 87mph at Wittering (Cambridgeshire), 85mph at Dover and 83mph at both Coltishall (Norfolk) and Cranwell (Lincolnshire). A Danish coaster sank off Cornwall with the loss of seven crew; there were two other gale-related deaths.

1978

3 January   *Tornadoes*
A swarm of tornadoes hit parts of East Anglia and Lincolnshire. Several minor injuries were reported, and there was some structural damage, notably near Newmarket.

11–12 January   *Severe gale; storm surge; heavy snow*
A severe northerly gale brought gusts to 89mph in central London (the Post Office Tower) and 83mph at Manston (Kent). A substantial North Sea storm surge resulted in coastal flooding from Yorkshire to Kent, and severe flooding followed in the Fen District

caused by the backing up of river water at high tide; Wisbech was worst hit. Snow accumulated to a depth of 20–25cm in Northern Ireland, and 30–40cm in eastern Scotland and north-east England. The death toll was 26.

25–29 January   *Severe snowstorm*
This was dubbed the 'great Highland blizzard'. The northern half of Scotland suffered its worst snowstorm since 1955, with many communities isolated for several days, dozens of motorists and train passengers stranded, and huge livestock losses. Helicopters were used for rescue work and delivering supplies. Undrifted snow depths of 40–60cm were common, and at Clashnoir, near Glenlivet, 90cm. The wind gusted to 87mph at Orlock Head in Northern Ireland. Four deaths were reported.

9–13 February   *Heavy snow*
Deep snow disrupted transport and cut power and telephone lines in eastern Scotland and north-east England. Depths reached 50–80cm in upland parts of Northumberland, the Scottish Borders and Aberdeenshire, and it lay 25–30cm deep in Newcastle, Edinburgh, Dundee and Aberdeen.

18–20 February   *Severe snowstorm; freezing rain*
A classic West Country blizzard. Worst affected were east Cornwall, Devon, Dorset, Somerset and much of south Wales. Villages were cut off for a week to ten days and helicopters were again used to deliver supplies. Undrifted snow depths included 70cm at Exton (Exmoor), 76cm at Widecombe (Dartmoor) and 85cm at Nettlecombe (Somerset), while drifts of 5 to 6m were widely reported. Seven people died. Freezing rain produced severe glaze on the 18th in the Channel Islands, Cornwall and parts of Devon, and on the 20th in south-east England.

10–11 April   *Heavy snow*
Heavy snow fell across the entire eastern half of the UK, with 20cm reported in several places.

15–16 June *Flooding*
Widespread flooding followed two days of torrential rain in parts of Yorkshire, Derbyshire and Nottinghamshire. A 48-hour rainfall total of 105mm was registered just west of Matlock.

16–17 September *Severe gale*
An intense depression containing the remnants of Hurricane 'Flossie' was responsible for a damaging gale in northern Scotland; Fair Isle logged 104mph and Kirkwall 100mph.

14 November *Severe gale*
Widespread gales affected Scotland, northern England and Northern Ireland. Fair Isle recorded a gust of 115mph, with 92mph at Kirkwall and also at Blackford Hill in Edinburgh.

27–28 December *Flooding*
Although there were no daily falls of note, between four and five times the normal December rainfall was recorded in part of north-east England, and the River Ouse rose 5m above its normal level at York where damaging floods occurred; 600 houses in the city were inundated. At Silent Valley (Co. Down) 214mm of rain fell in 48 hours and at Arkengarthdale (Co. Durham) there were 108mm in the same period.

31 December to 4 January 1979 *Snowstorms*
Dry powdery snow was driven into deep drifts by an easterly gale. Estimated undrifted depths of 20–30cm were noted in eastern Scotland, north-east England and in parts of southern England, notably over Dartmoor. Drifts 2 to 4m high were reported. Drifted snow halted air traffic at Heathrow for several days. A fall of 25cm affected north Staffordshire and adjacent areas on the 2nd, and a blizzard swept Cornwall, Devon and the Channel Islands on the 4th; Jersey was paralysed for several days, with level snow 30cm deep and drifts to 5m.

1979

19–23 January   *Heavy snow; freezing rain*
A belt of heavy snow moved northwards across the country followed in southern districts by a temporary thaw, but in northern England 40–50cm accumulated in upland districts. Heavy snow returned to southern and central parts on the 22nd–23rd, and falls of 15–20cm were followed by a period of freezing rain which caused chaos on roads, railways and the London Underground.

13–16 February   *Snowstorms; storm surge*
On the 13th a storm surge and exceptionally heavy swell caused flooding at a number of places along the English Channel coast; Chesil Beach was breached and the Isle of Portland cut off for a time. Widespread heavy snow followed between the 14th and 16th, and severe drifting isolated hundreds of towns and villages, including for a time Norwich, Ipswich and Felixstowe. Undrifted depths of 25–30cm were typical in East Anglia and the East Midlands, with 50–60cm in upland parts of Derbyshire and Yorkshire. Drifts to 6m were noted.

16–19 March   *Heavy snow*
Widespread heavy snow fell across large parts of the UK during this period with 15–25cm of level snow widely reported. At Shawbury (Shropshire) and at Durham the depth was 31cm, while at Gosforth, a suburb of Newcastle-upon-Tyne, the depth was 45cm. Both Newcastle and Aberdeen were inaccessible for a time, and hundreds of villages were isolated.

2 May   *Heavy snow*
An exceptional late snowfall deposited 20–30cm in upland parts of Devon, Somerset and Wales.

13–14 August   *Severe gale*
Often dubbed the 'Fastnet storm', and sometimes the 'Fastnet fiasco', this exceptional summer gale wreaked havoc during the Fastnet race, and took the lives of fifteen competitors; a further ten weather-related deaths were reported at sea and on land on the same date. Peak gusts included 75mph at Milford Haven (Pembrokeshire) and 85mph at Hartland Point (Devon).

4–15 December   *Severe gales*
A sequence of destructive gales swept the country, uprooting thousands of trees, causing much structural damage and resulting in several shipping losses in British coastal waters; the death toll was twenty. On the 17th a radio transmitter at Droitwich collapsed. Peak gusts were 110mph at Blackford Hill, Edinburgh, on the 4th, and 118mph at Gwennap Head, Cornwall, on the 15th.

26–27 December   *Severe flooding*
Prolonged heavy rain affected most parts of the UK and flooding was reported widely. Worst hit were southern England and south Wales, and at Venford reservoir, above Ashburton (Devon), 223mm of rain was measured over the 48-hour period. In south Wales four people were drowned. Approximately 2,000 sq. km received 100mm or more.

1980

5 June   *Tornado; thunderstorms; flooding*
A day of widespread thunderstorms and localized downpours; in Darwen (Lancashire) flooding followed a deluge of 67mm in less than two hours; at Nairn in north-east Scotland a tornado caused extensive damage to a caravan site; across the country at least three people were killed by lightning.

25 June   *Thunderstorm; heavy hail*
At Sevenoaks School 116mm of rain fell in 105 minutes; hail lay 25cm deep in the vicinity.

26 July   *Thunderstorms; flooding*
Severe thunderstorms and associated flooding affected a wide area of southern and central England; 98mm of rain fell at Brixworth and 100mm at Pitsford, both in Northamptonshire.

1 August   *Flash flood; heavy hail*
A phenomenal downpour dropped 97mm in 45 minutes at Orra Beg on the Antrim plateau near Cushendun. Hailstones 5cm across were observed. It is very likely that appreciably more than 100mm fell in less than an hour at the centre of the storm. Much landscape damage ensued.

7 August   *Flooding*
Flooding in the Wells/Fakenham district of north Norfolk was triggered by falls of 105mm at Cockthorpe and 103mm at Houghton St Giles, most of the rain falling in five hours.

14–15 August   *Flooding*
Widespread flooding, notably in Northampton, Kettering, Bedford and Luton, followed a night of torrential rain. Weekley, near Kettering, recorded 107mm and Higham Ferrers 99mm.

29 September and 10 October   *Flooding*
A sizeable area of West Sussex centred on Worthing suffered serious flooding twice in a fortnight; in Worthing itself 95mm fell on the 29th and 101mm on the 10th.

26–27 October   *Flooding*
Floods affected rivers draining Snowdonia following a fall of 213mm in 48 hours at Blaenau Ffestiniog.

1981–90

1981

22–23 February   *Heavy snow*
Heavy drifting snow affected Wales and the western Midlands; 35cm accumulated at Hednesford (Staffordshire).

9–10 and 21–23 March   *Severe flood*
Two instances of prolonged orographic rain, heaviest in south Wales and south-west England, triggered widespread flooding in these regions and also in the Midlands. West Country trains could get no further than Exeter thanks to floodwater. Over 200mm of rain fell inside 48 hours on both occasions. Monthly totals approached 800mm in upland parts of Glamorgan and Carmarthenshire.

14 April *Thunderstorm*
In arguably the most intense April thunderstorm on record, 86mm fell in less than eight hours at Horsham (Sussex).

24–26 April *Severe snowstorm; flooding*
Three days of heavy snow affected large parts of the UK, and although accumulations at low levels were mostly small, undrifted snowdepths of 30–50cm were typical above about 150m above sea-level. At Berry Hill in the Forest of Dean snow lay 66cm deep, and at Middleton in Derbyshire 59cm deep. Drifts of 6m or more were reported. In Lincolnshire and Norfolk prolonged heavy rain – 126mm in 72 hours at Sheringham – caused serious flooding; Horncastle was very badly hit. A rapid thaw also led to flooding in the Midlands and Yorkshire. Seven weather-related deaths were reported.

9 July *Thunderstorms; flooding*
Severe thunderstorms across southern England; London badly affected. 104mm of rain fell in ten hours at Heronsgate, near Brentwood in Essex. A woman was killed by lightning in Somerset.

5–6 August *Thunderstorms; flooding*
Widespread thunderstorms led to flooding in Greater Manchester and Cheshire, in Northamptonshire and in London. At Eaton, near Tarporley (Cheshire), 132mm of rain was recorded, and the same amount was logged at Norton Junction, near Daventry, Northamptonshire.

23 November *Tornadoes*
The UK's biggest single outbreak of tornadoes affected many places in northern England, Wales, the Midlands and East Anglia. Structural damage and minor injuries were reported. The tornado count was eventually put at 105.

7–28 December *Heavy snowfalls; severe cold; gales; storm surge*
This was the snowiest December of the twentieth century, with major falls across central and southern regions on the 8th, 11th, 13th–14th, 20th–21st and 24th. Snow depth reached 26cm at Heathrow on the 11th, 31cm at Shrewsbury on the 14th, 33cm at

Lincoln on the 22nd and 91cm above Hawes, Yorkshire, on the 14th. On the 13th a severe gale caused damage in the West Country and an associated storm surge resulted in coastal flooding around the Bristol Channel; a gust of 89mph was recorded at Burrington, Devon. A further gale on the 19th was responsible for the Penlee lifeboat disaster off the Cornish coast in which sixteen perished. In all there were 30 weather-related deaths this month, but the estimated death toll resulting from hypothermia was between 500 and 1,000; the temperature dropped to −25.2°C at Shawbury (Shropshire) on the 13th.

1982

2–8 January *Widespread flooding*
Edinburgh airport collected 110mm of rain between the 2nd and 5th and similar figures were obtained throughout the Lothians, bringing about a good deal of flooding in the district. There was also severe flooding in Yorkshire between the 5th and 8th, described as the worst since 1947, caused by a combination of heavy rain and a rapid thaw, although a dramatic drop in temperature after the 5th led to the floodwaters freezing over. At York, the Ouse rose to 5m above normal, and the York, Selby and Tadcaster districts were very badly hit.

4–15 January *Snowstorm; freezing rain; severe cold*
The first major snowfall of this spell affected eastern and central Scotland and north-east England, with 15–25cm falling over a wide area, and 40cm at Braemar, Aberdeenshire. The blizzard of the 8th–9th affected most of southern England, Wales and the south Midlands, and the snow drifted readily in the gale-force easterly wind at temperatures typically between −2 and −6°C. Some 30–50cm fell widely, with drifts to 6m, but in upland parts of Glamorgan and Breconshire undrifted depths approached 100cm. Many towns and cities, including Cardiff and Swansea, were cut off for a time, and some remote villages were isolated for up to two weeks and suffered lengthy power failures. The Territorial Army was used to deliver supplies. Heavy freezing rain produced extensive glazed ice on the 9th in Devon, Somerset and Dorset, and on the 10th in the Channel Islands. There followed a period of extreme cold, and the

national temperature record of –27.2°C was equalled at Braemar on the 10th while daytime maxima between –10 and –15°C occurred on several days. The death toll from hypothermia was conservatively estimated at 1,000.

3 March *Severe gale*
Some structural and landscape damage, as well as dislocation of traffic and disruption of power supplies, followed a severe gale over Scotland, northern England and Northern Ireland. A ship broke up off Stranraer. Gusts to 100mph were recorded at Blackford Hill, Edinburgh, and St Abb's Head, Berwickshire.

1–23 June *Thunderstorms; heavy hail; severe flooding*
The first nine days of this month were hot and humid with severe local thunderstorms leading to serious but localized and short-lived flooding. On the 2nd, 90mm of rain fell in two hours at Wootton Bassett, near Swindon; on the 4th, 92mm fell, mostly in 70 minutes, at Cheshunt, Hertfordshire; and on the 6th, 96mm fell at Skipton, Yorkshire, in five hours. On the 18th a severe rain and hailstorm hit the Bristol and Bath districts; 83mm of rain fell at Stanton Prior, mostly in two hours, and piles of hail collected to a depth of 60cm in the Speedwell district of Bristol. A prolonged downpour on the 21st and 22nd caused destructive and long-lasting floods in Yorkshire and north Derbyshire; over the two days 114mm of rain fell at Brockholes, above Sheffield, 103mm at Weston Park in Sheffield itself and 110mm at Chesterfield. Leeds, Sheffield, Doncaster, Rotherham and Chesterfield were all very badly hit, while the failure of embankments along the trans-Pennine M62 resulted in piles of mud 3m high blocking the carriageways. The total area receiving 100mm or more during these two days was approximately 1,200 sq. km.

11–12 July *Thunderstorms; flash flood*
Several hours of torrential rain accompanied vivid thunderstorms in Devon, Dorset and Somerset, leading to serious flooding. A flash flood in the small Somerset town of Bruton caused extensive damage. Upton Noble, near Bruton, collected 113mm of rain in sixteen hours, while Burrington in Devon logged 91mm.

1–7 August *Thunderstorms; flooding*
Storms broke out widely during this very humid week, triggering some severe local floods. At Ely, 124mm fell in three hours on the 4th, and at Blairgowrie, Perthshire, 109mm fell on the 6th.

15–20 December *Severe gales*
Two destructive gales swept much of the UK with extensive damage to trees, buildings, power and telephone lines, and disruption of transport. In Leeds many public buildings (as well as private homes) were seriously damaged. Two ferries collided on the 19th off Harwich, with six dead.

1983

31 January to 1 February *Severe gale; storm surge*
A west to north-westerly gale blasted across the entire country with highest gusts including 98mph at Sumburgh (Shetland), 94mph at Lynemouth (Northumberland) and 93mph at Sunderland. Structural and woodland damage were widely reported, and transport badly disrupted. A North Sea storm surge led to a limited amount of coastal flooding from Northumberland to Norfolk, and the newly commissioned Thames Barrier was raised for the first time.

8–11 February *Heavy snow*
Several falls of snow affected eastern Scotland, eastern England and parts of the Midlands during this period; 50cm lay at Braemar, 36cm at Dover and 30cm at Folkestone.

4–6 March *Flooding*
The extreme north of Scotland experienced considerable flooding following a three-day downpour. The fall of 349mm recorded at Gobernuisgach Lodge, Sutherland, was a 72-hour record for the UK.

3 April *Heavy snow*
Easter Day travel was badly disrupted by widespread snowfalls; 20cm lay in east Kent.

27–28 May   *Flooding*
Extensive and destructive flooding in the Scottish Borders and north Northumberland followed a prolonged downpour in which 137mm fell in 48 hours at Hungry Snout (West Lothian), and 100mm in 24 hours at Sourhope (Roxburghshire).

5–7 June   *Thunderstorms; heavy hail; flooding*
Violent thunder and hailstorms hit south coast counties of England on the 5th; at Winfrith (Dorset) 74mm of rain fell, mostly in two hours, along with hailstones 7cm across. Much damage to glasshouses and slate roofs followed, and there were even reports of yachts at Christchurch capsizing under the weight of the hail. Similar sized hailstones were reported from the Manchester area on the 7th.

23 June   *Thunderstorms; flooding*
Mid- and east Hampshire and adjacent parts of Surrey and Sussex were worst hit; 103mm of rain fell at Rotherfield Park and 93mm at Brown Candover.

6 July   *Thunderstorms; flooding*
Storms broke out widely across England; 95mm of rain fell at Idle Hill, near Sevenoaks.

17 July   *Severe flooding*
Destructive floods hit the northern Pennines following a torrential downpour, and there were reports of peat-bogs bursting in west Durham. At Honister Pass, Cumbria, 112mm of rain fell, with 104mm at Ireshopeburn in Upper Weardale and 98mm at Appleby Castle.

July   *Heatwave*
The hottest July in over 300 years of records (subsequently eclipsed by July 2006) led to over 500 deaths from heatstroke and dehydration, mostly among the elderly. During the peak of the heatwave, between the 4th and 18th, the death rate among the over-75s rose by 50 per cent.

2–5 September   *Severe gales*
Exceptionally severe gales for so early in the season swept much of
the UK causing great losses in the fruit-growing areas of Worcestershire and Kent. Power lines were brought down, there was some
structural damage and four people died. Peak gusts were 86mph at
Gwennap Head (Cornwall) and 84mph at Dover, both on the 2nd.

20 December to 13 January 1984   *Severe gales*
The gale on the 20th coincided with spring tides, and a storm surge
in the English Channel caused flooding in a large number of places
along the south coast of England. The gale of the 31st was widespread and damaging with a maximum gust of 106mph at Lynemouth. High winds on the 3rd were confined to central and
northern Scotland, but included 103mph at Benbecula (Western
Isles) and Duirinish (Wester Ross). The gale of the 13th was the
most widespread with structural damage reported from many parts
of the UK, including the collapse of a cooling tower at Widnes;
maximum gust was 99mph at Blackford Hill, Edinburgh, and
further south a noteworthy 95mph was recorded at Hemsby in
Norfolk. On the 14th a tornado damaged 100 homes in Doncaster.
At least twenty people died in these gales.

1984

18–27 January   *Snowstorms*
Heavy snow fell repeatedly over Scotland, Northern Ireland and
northern England, with massive accumulations in the northern
Pennines, the Cheviot Hills, the Southern Uplands and the southern
flank of the Scottish Highlands. At times the Central Lowlands were
badly hit, too. Undrifted snow was 112cm deep at Leadhills (Lanarkshire), 68cm deep at Eskdalemuir and 58cm deep at Braemar. Four
trains were trapped in drifts on the 21st–22nd, 2,500 skiers were
marooned at Glenshee and most had to sleep in their vehicles, and
dozens of motorists were rescued by helicopter. At least seven people
died from exposure during this period.

9 July   *Lightning*
Thunderstorms broke out widely on this date. The thirteenth-

century north transept of York Minster was badly damaged by fire following a lightning strike.

23–24 July   *Thunderstorms; flooding*
Torrential downpours associated with severe thunderstorms affected southern counties of England. The Portsmouth, Chichester and Horsham areas were worst hit, with 92mm of rain at Hayling Island and 87mm at Havant.

July and August   *Drought*
An acute water shortage affected primarily south-west England, Wales and Northern Ireland as the result of a seven-month (February to August) drought which had only twice been exceeded, in 1959 and 1976, in the preceding 100 years. However, this was also the earliest known instance of blatant (as distinct from subtle) 'spinning' of the facts: water authorities invited the media to view a dried-out reservoir in Devon which had in fact been emptied for maintenance purposes.

18 October   *Widespread gales*
Damage to buildings, power lines and woodlands occurred extensively, and gusts included 115mph at Pendennis Point (Cornwall) and 106mph at Salsburgh (east of Glasgow).

3 November   *Severe flooding*
Widespread floods affected central and eastern Scotland and north-east England following falls of 110mm at West Hopes reservoir (East Lothian), 105mm at Tillicoultry (near Stirling), 100mm at Whitchester (Berwickshire) and 98mm at Whitehillocks (Angus).

1985

5–22 January   *Frequent snow*
Several moderate snowfalls, chiefly in eastern districts, caused repeated disruption to traffic. Some 20–25cm of undrifted snow affected eastern parts of Kent, Essex and Suffolk, with smaller falls elsewhere. A general thaw set in on the 19th, but northern Scotland had a snowstorm on the 22nd which left 50cm on the ground at Aviemore.

7–12 February   *Drifting snow*
Snow fell widely on the 8th–9th in central and southern parts of England and in Wales, with undrifted depths of 20–30cm in mid- and north-east Wales and parts of the western Midlands. Strong east winds caused persistent drifting between the 10th and 12th.

June to September   *Poor summer*
Poor ripening conditions and waterlogged fields at harvest-time characterized a summer that was disastrous for farmers in many parts of Scotland, Northern Ireland and northern England. In the Glasgow area it was the wettest summer for over 100 years, and the coldest for over 50.

26–29 December   *Heavy snow*
Snow fell widely but was heavy only in north-east England where depths of 20–30cm occurred.

1986

7–8 January   *Heavy snow*
Widespread snow over England and Wales was heaviest in the north Midlands where Hulland (Derbyshire) reported 38cm.

28–30 January   *Heavy snow*
A narrow zone of heavy snow affected chiefly land above 150m in Wiltshire, Gloucestershire, Herefordshire, Shropshire, Staffordshire, Derbyshire and Yorkshire. Some 35cm fell at Holme Moss, Clee Hill and Ruardean in the Forest of Dean. The snow was soft and clinging, and brought down hundreds of branches and many kilometres of power and telephone lines.

February   *Severe cold; freezing rain; heavy snow*
Bitterly cold easterly winds blew throughout this month which was the coldest February, that of 1947 apart, of the twentieth century. Deaths from hypothermia were estimated at over 1,600. Freezing rain caused many problems during the first three days of the month in upland parts of northern England, the Midlands and Wales, and over the highest parts of the Cotswolds and Chilterns. There was

generally little snow, except on the 6th–7th when 15–25cm fell in parts of the Home Counties, Hampshire and Isle of Wight.

24 March   *Severe gale*
A narrow zone across south Wales and the south Midlands experienced widespread damage. Highest gust was 93mph at Cardington, Bedfordshire. Power failures affected thousands of homes in Oxfordshire and the northern Home Counties.

25–26 August   *Severe flooding; gales*
A depression containing the remains of Hurricane 'Charley' caused widespread chaos on this August bank holiday. The geographical area affected by 25mm and 50mm of rain established new records for both thresholds, and more than 100mm fell in several different parts of Wales, and also in County Durham and in Yorkshire. The small town of Whitland, 25km west of Carmarthen was completely inundated, to a depth of 2m in some parts. Highest 24-hour rainfall totals were 135mm at Aber (east of Bangor), 121mm at Walshaw Dean (near Halifax) and 109mm at Loggerheads (near Wrexham). More than 100mm of rain fell over an area of approximately 1,500 sq. km, taking the two days together. Winds were also unusually strong for August, with gusts to 75mph at Brixham (Devon) and 71mph at Gwennap Head (Cornwall). At least eleven people died in the floods and gales.

21 November   *Tornado*
A strong tornado tore through the Sussex town of Selsey, damaging 250 houses.

29 December   *Severe flood*
Destructive floods hit Dolgellau, Blaenau Ffestiniog and Machynlleth, and road and rail transport in mid- and north Wales were badly disrupted. Aberangell recorded 158mm and Machynlleth 142mm.

1987

7–20 January   *Heavy snow; severe cold*
This cold spell, although quite short, was probably the most intense

of the entire twentieth century. On the 12th, maximum temperatures were widely between –6 and –8°C, and at Warlingham, Surrey, the mercury climbed no higher than –9.1°C; St Mary's (Isles of Scilly) recorded a minimum of –7°C the following night accompanied by a 40mph north-easterly wind. The national death toll from hypothermia was estimated at approximately 500. Exceedingly heavy snow fell between the 11th and 14th, but over a limited area; maximum snow depths included 75cm over the North Downs just east of Maidstone, 55cm at Southend, 52cm at East Malling and 48cm at Gillingham (both Kent), 46cm at Cottesmore (Rutland) and 45cm at Orpington (south-east London). West Cornwall was also badly hit, with 39cm at Penzance and 30cm at Falmouth.

19 March   *Heavy snow*
A narrow zone of heavy snow extended from south Gloucestershire to the Isle of Wight and West Sussex; accumulations of 25–30cm were noted around Bath, Romsey and Cowes.

27 March   *Severe gale*
A destructive gale swept southern England and south Wales, with gusts to 107mph at Gwennap Head, 100mph at Burrington (Devon) and 98mph at Rhoose (Glamorgan).

23 August   *Thunderstorms; severe flooding*
Widespread thunderstorms occurred between the 21st and 23rd. On the 23rd severe local flooding was triggered by falls of 116mm at Holbeach (Lincolnshire) and 115mm at Blithfield reservoir (east Staffordshire), but serious flooding also affected the Wigan area and parts of Essex.

9–10 October   *Severe flooding*
Some 50–80mm of rain fell widely across the western Home Counties from Surrey to Bedfordshire, and also in Hampshire and the Isle of Wight. 90mm was recorded at Totland Bay. Many roads and railways were impassable for a time.

15–16 October   *Destructive gale*
The most destructive gale of the century, and possibly since 1703, battered England south-east of a line from Dorset to The Wash,

damaging thousands of homes and public buildings, uprooting an estimated 15 million trees, disrupting power to practically the entire Home Counties including most of London, and killing eighteen people. At Shoreham, Sussex, the highest mean hourly wind was 85mph with a peak gust of 115mph. Other exceptional gusts were 108mph at Langdon Bay, near Dover; 106mph at Ashford, Kent; 103mph at Thorney Island, near Chichester, and Herstmonceux, near Eastbourne; and 100mph at Shoeburyness, in Essex.

18–19 October *Severe flood*
A railway bridge collapsed in Carmarthenshire and a train plunged into a flooded river, killing five people. At Trecastle, Breconshire, 222mm of rain fell in the five-day period ending on the 18th.

1988

9 February *Severe gale*
Road and rail transport were disrupted and trees uprooted in many parts of the UK; structural damage was worst in the West Country and in north-west England where the roof of Bury town hall was badly damaged. Gusts included 110mph at Gwennap Head, 106mph at Hazelrigg (near Lancaster) and 90mph at Culdrose (Cornwall).

8–9 May *Thunderstorms; flooding*
Flooding followed thunderstorms in west London and the Thames Valley. Ruislip recorded a total of 89mm of rain in two separate storms.

19 October *Thunderstorms; flooding*
Severe flooding in Liverpool and the Wirral during a dramatic thunderstorm which dropped 82mm, mostly in 75 minutes, at Crosby.

20 November *Heavy snow*
A belt of snow moved southwards across the country; east Kent had 15–20cm, resulting in Dover being cut off for a time. Over 30cm fell over high ground in north Wales and the Isle of Man.

1989

5–6 February   *Severe flood*
The 48-hour long downpour was exceptional even by the standards of the Western Highlands; 215mm was recorded at Fort William, 261mm at South Laggan (on Loch Oich), 285mm at Clunes Forest (Loch Lochy), and close to 300mm at Kinlouchhourn (on Loch Hourn). More than 100mm fell over an area of almost 10,000 sq. km. Destructive floods hit large parts of the Highlands, including the length of the Great Glen and the Inverness and Dingwall districts. In Inverness, the rail bridge over the River Ness, built in 1862, was swept away.

13 February   *Severe gale*
Gales raked the whole country, causing damage from the north Midlands northwards; the winds were especially violent across northern Scotland where much landscape and structural damage ensued. Kinnaird Head, near Fraserburgh, reported a record low-level gust for the UK of 142mph, with a mean hourly speed of 76mph. Other notable gusts were 115mph at Fair Isle, 107mph at Benbecula and 106mph at Butt of Lewis.

5 April   *Heavy snow*
A heavy fall of snow disrupted traffic in southern England; 18cm lay at Tadworth (Surrey). Elsewhere, the high-level site of Holme Moss, West Yorkshire, had 30cm.

11 April   *Severe gale*
Wales was worst affected by this gale, with gusts to 97mph at Milford Haven and 90mph at Aberporth.

19 May   *Flash flood*
At Walshaw Dean reservoir, above Halifax, 193mm of rain fell in two hours. The resulting flash flood was destructive in Hebden Bridge, Mitholmroyd, Sowerby Bridge and the outskirts of Halifax.

24 May   *Thunderstorm; flooding*
Widespread thunderstorms and torrential rain affected many parts of southern England; at Swallowcliffe, west of Salisbury, 110mm of rain was recorded, mostly in three hours.

11 September *Thunderstorm; flooding*
Torrential rain hit a sizeable part of south Devon triggering severe local flooding; 123mm fell at Holsome, 109mm at Avonwick and 98mm at Slapton Ley.

28 October *Severe gale*
Southern coastal counties were swept by damaging winds gusting to 101mph at Portland Bill.

14–21 December *Gales; flooding*
Repeated bouts of heavy rain, amounting to 160mm in ten days in Hampshire and parts of neighbouring counties, led to extensive flooding. Frequent gales exacerbated the situation, bringing down hundreds of trees, while high tides on the 16th led to coastal flooding, notably in Plymouth.

1990

25 January *Destructive gale*
Most of Scotland missed the extreme effects of the so-called 'Burns' Day storm'. Over England and Wales, however, it was arguably the most severe gale of the twentieth century for geographical extent and intensity combined, and the death toll of 47 certainly supports that contention. Structural damage was reported from all parts of England and Wales, tens of thousands of trees were uprooted, transport was severely disrupted, dozens of high-sided vehicles were blown over and millions lost electricity and telephone services. The gale lasted 15–18 hours, and highest recorded gusts were 107mph at Aberporth (Ceredigion), 107mph at Gwennap Head (Cornwall), 102mph at Culdrose (Cornwall), 101mph at Sheerness (Kent), 98mph at St Mawgan (Cornwall), 98mph at Herstmonceux (Sussex) and 97mph at Shoreham (Sussex).

February, especially the 26th *Frequent severe gales; coastal flooding*
This was probably the windiest month of the century, and the sequence of severe gales which began on 25 January continued until 5 March. On 11 February a gust of 93mph was recorded at Burrington (Devon) and structural damage was reported from many parts

of the southern England, but the 26th was the wildest day with gusts of 100mph at St Abb's Head (Berwickshire), 99mph at Hemsby (Norfolk), 98mph at Leeds and 92mph at Fylingdales (both Yorkshire). Eleven people died on this day. The north-westerly gale combined with high spring tides to produce a storm surge along southern and eastern coasts of the Irish Sea, and the sea wall at Towyn (Denbighshire) failed catastrophically, resulting in the flooding of several hundred homes.

14–15 March   *Flooding*
Flooding returned to many glens in the Scottish Highlands, including the Inverness area, just thirteen months after the catastrophic floods of February 1989. On this occasion 240mm of rain fell at Kinlochhourn, of which 187mm fell on the 15th alone.

1–4 August   *Heatwave*
Four days with maxima widely in excess of 32°C resulted in several hundred deaths, mostly among the elderly, from heatstroke and dehydration. The temperature reached 37.1°C at Cheltenham, breaking a record which had stood for 79 years, but which was to be broken again in 2003. Overnight minima between 20 and 23°C were reported from several sites along the south coast and in London.

28 October   *Severe gale; coastal flooding*
A gust of 99mph was recorded at Herstmonceux (Sussex) during this gale, the effects of which were most seriously felt in southernmost counties of England. A total of thirteen people died in road accidents in which the bad weather was implicated. High tides and a storm surge combined to cause flooding at several sites along the English Channel coast.

7–8 December   *Heavy snow*
Snow fell heavily for up to eighteen hours across much of England and Wales, and undrifted depths of 20–30cm were reported from many parts of the Midlands, Wales and Yorkshire. In the Birmingham area 25–40cm was typical, with 42cm at Acocks Green. A rainfall equivalent of 97mm was recorded at Fylingdales (North Yorkshire) while the northerly wind gusted to 85mph at Warcop (Cumbria). The Midlands motorway network quickly became grid-

locked, and many drivers were stranded for up to 48 hours (see Chapter 2). The snow was wet and clinging, and brought down overhead power and telephone wires. Six people died.

25 December   *Tornadoes*
A swarm of tornadoes caused mainly minor structural damage at several locations throughout England and Wales.

1991–2000

1991

1–8 January   *Severe gales*
The weather was very rough during this period, especially on the 5th when a gust of 90mph was recorded at Greenock, and storm surges caused a good deal of flooding around western and southern coasts of the UK, from the Firth of Clyde in the north to Sussex in the south-east.

7–9 February   *Heavy snow*
Snow fell, often heavily, for 48 hours over large parts of England and Wales, with maximum depths of 51cm at Wilsden (West Yorkshire), 46cm at Longframlington (Northumberland), 45cm at Fylingdales (North Yorkshire), 35cm at Pencelli (Breconshire) and 30cm at Honington (Suffolk). The snow was dry and powdery, and seriously infiltrated the hydraulic door-closing mechanism on many new trains, resulting in the immortal 'wrong kind of snow' line uttered by a British Rail spokesman. Further snow fell around the 13th before a slow thaw set in.

13 March   *Thick fog*
Thick fog during the morning rush hour was implicated in a pile-up on the M4 in which ten people died.

28–29 September   *Flooding; gales*
Two days of very wet and windy weather caused major problems for transport by road and rail, and at sea. All ferry crossings to Ireland

were suspended for a while. Flooding was particularly bad in Hampshire, Dorset, Wiltshire and Somerset, and 114mm of rain fell at Branksome and 110mm at Poole (both Dorset), 106mm at Currymoor and 93mm at East Lyng (both Somerset), and 96mm at Velindre (Breconshire).

12 November   *Severe gale; tornado*
Disruption to transport and some structural damage was reported as winds gusted to 93mph at Aberporth (Ceredigion) and 91mph at Ronaldsway (Isle of Man). A tornado struck the Cambridgeshire village of Dullingham.

20–21 December   *Flooding*
Hundreds of homes were flooded in the Greater Manchester and Sheffield areas, among others, following a 48-hour downpour which dropped 100–150mm of rain over the southern Pennines. On the 21st alone 117mm fell at Holme Moss (West Yorkshire).

31 December to 1 January 1992   *Severe gale*
A destructive gale swept the extreme north of Scotland, and damage was very extensive in Shetland where the domes at the early-warning station at Saxa Vord were demolished. Highest gusts included 113mph at Sumburgh, 110mph at Butt of Lewis, 109mph at Lerwick, 108mph at Fair Isle, 107mph at Kirkwall, and 103mph at Sella Ness.

1992

31 March to 2 April   *Severe flood*
Prolonged heavy rain led to widespread flooding on rivers in south-east Scotland and north-east England. Urban flooding was extensive on Tyneside, especially in Newcastle where 110mm of rain fell in 60 hours.

29 to 31 May   *Severe thunderstorms; flooding*
Three days of widespread and often severe thunderstorms, accompanied by torrential downpours, led to serious local flooding, notably in Hertfordshire, Bedfordshire, Buckinghamshire and London. At Hitchin 112mm of rain was recorded in two separate storms on the 29th, and 133mm fell in two hours at Chorleywood on the 31st.

29–31 August   *Gales; heavy rain*
This exceptionally windy and wet August bank holiday weekend disrupted many thousands of people's plans for a few days on the coast; indeed, many seaside promenades were closed. The death toll was six. Gusts included 82mph at Fair Isle and 68mph at Sheerness, Kent.

18–23 September   *Severe flooding*
On the 18th a violent thunderstorm in Wiltshire dropped 98mm of rain at Upton Scudamore, near Warminster. Prolonged heavy rain on the 22nd–23rd led to disruptive floods in the north London suburbs, and also along the floodplains of the Great Ouse and Nene in Bedfordshire, Northamptonshire and Cambridgeshire. At Riseley (Bedfordshire), 106mm of rain fell in 48 hours.

29 November to 2 December   *Severe flooding*
The worst flooding since 1979 hit the Welsh Valleys, and serious floods also affected many parts of south-west England, including Bristol, and the floodplains of the Severn, Wye and Thames. At Treherbert, Glamorgan, 250mm of rain fell in 96 hours.

15–16 December   *Heavy snow*
Heavy snow blocked roads in central and north-eastern Scotland; level depths included 45cm at Aviemore, 29cm at Glenlivet and 16cm at Inverness.

1993

4–24 January   *Severe gales; storm surge; snowstorms; severe flooding*
January 1993 rivalled February 1990 as the windiest month of the century. A sequence of destructive gales swept the UK between the 4th and 24th; the oil-tanker *Braer* was driven onto rocks in Quendale Bay, Shetland, on the 5th, broke up progressively in rough seas over the next few days and finally sank on the 12th; damage to buildings, woodland and power lines occurred widely and frequently, and many people lost their lives. The gale of the 13th was particularly damaging in southern counties, but the highest gusts occurred on the 17th, with 108mph at Kirkwall, 107mph at Sumburgh and 106mph at Shin (Sutherland). Tiree recorded

97mph on the 21st and Leeds 95mph on the 24th. A storm surge associated with the gale of the 10th caused flooding in a large number of places around southern and western coasts of the UK. On the 4th, 25–35cm of snow disrupted transport in West and South Yorkshire and north Derbyshire, and a combination of heavy wet snow 25–50cm deep and gale force winds caused chaos in central and southern Scotland on the 11th–12th. 51cm of snow lay at Aviemore. Heavy rain and a rapid thaw led to severe flooding in many parts of Scotland between the 15th and 18th, nowhere worse than on the Rivers Earn and Tay; at Perth the flood was reported to be the highest since 1814.

21 February   *Severe gale; storm surge*
A north-westerly gale affected especially northern and eastern districts, triggering a North Sea storm surge which led to serious local flooding at several locations between the Humber and the Thames. A gust of 100mph was recorded at Sumburgh (Shetland).

26 February to 1 March   *Heavy snow*
Snow disrupted transport in northern and eastern Scotland and north-east England; an undrifted depth of 30cm was reported from the Huntly district of Aberdeenshire.

13–15 May   *Flooding; heavy snow*
At Dungonnell (County Antrim) 97mm of rain fell, with 75–80mm in parts of southern Scotland and north-east England, leading to some flooding. Some 20–30cm of snow fell in the Southern Uplands above about 500m.

26 May   *Severe thunderstorm; flooding*
West Berkshire and west Oxfordshire were worst affected by a violent thunderstorm which dropped 129mm, mostly in three hours, at Uffington. Two other sites recorded over 100mm.

9 to 11 June   *Thunderstorms; severe flooding*
A nine-hour-long deluge at Culdrose, Cornwall, amounted to 123mm and created a destructive flash flood in nearby Helston. The next day severe storms in Essex delivered 121mm to North Weald and 92mm to Thornwood Common, most of it falling in 150

minutes, and severe flooding hit Harlow, Epping and Ongar. On the same day, Llandudno collected 175mm, mostly falling inside four hours, and major flooding was reported in many parts of Wales. A prolonged downpour on the 11th–12th dropped 151mm at Aberporth and 162mm at Davidstow Moor, and the flood at nearby Bude was reckoned to be the worst since 1904.

12 September to 13 October *Frequent flooding*
This exceptionally wet period was noteworthy for the repeated episodes of flooding in eastern, central and southern England. On 12 September, Devon was badly hit, with 114mm at Swincombe and 105mm at Fernworthy; between 13th and 16th September 159mm fell in 84 hours at Fylingdales (North Yorks); Gatwick Airport collected 90mm in 48 hours beginning on the morning of 30 September; and between the 10th and 12th October 128mm fell in 72 hours at Theberton in east Suffolk and 96mm at nearby Aldeburgh.

8–9 December *Severe gale*
Structural damage was reported widely across southern and central parts of England and Wales; fourteen people died. Highest gusts were 97mph at Camborne and 95mph at Aberporth. This was probably the most widely damaging gale since 1990.

30 December to 10 January 1994 *Widespread flooding*
Serious flooding again hit many parts of southern England. There was a flash flood at Polperro, Cornwall, on the 31st following an unmeasured deluge to the north of the village; Chichester was virtually cut off for several days in early January when the River Lavant overtopped its banks; several houses collapsed in Haywards Heath when their foundations were washed away; 170mm of rain fell in two weeks at Southsea. There was flooding, too, along the Great Ouse, the Thames and in Northern Ireland.

1994

23 January *Severe gale*
Northern Scotland was battered by a severe gale with gusts to 120mph at Sumburgh and 109mph at Lerwick.

26 February to 1 March   *Snowstorm; flooding*
Snow lay 30cm deep at Aviemore on the 1st, but as much as 50–60cm may have fallen in the hills of Aberdeenshire. Major flooding in Aberdeen followed 118mm of rain in 72 hours.

1 April   *Severe gale*
Widespread damage, notably at the resort of Barry Island, accompanied a severe gale in Wales and south-west England. Cardiff recorded a gust of 100 mph.

24 July   *Thunderstorms*
Lightning struck the oil terminal at Milford Haven, triggering explosions and a fire. Thunderstorms broke out widely that day, and 74mm fell in a few hours at Denstone (Staffordshire).

4 August   *Thunderstorms*
A night of widespread storms; 100mm of rain was recorded at Capheaton (Northumberland).

31 August to 1 September   *Flooding*
Severe flooding followed heavy rain in parts of Norfolk and Suffolk; 146mm of rain fell overnight at Ditchingham, near Bungay.

6–12 and 26–28 December   *Severe flooding*
Extended incursions of moist tropical air led to continuous heavy rain over south-west-facing uplands. During the first episode, 170mm of rain was recorded at Loch Sloy on the 10th alone, with 350mm in the week between the 6th and 12th; devastating floods affected large parts of the Central Lowlands and the Southern Highlands, and Glasgow, Paisley and Kirkintilloch were particularly badly hit. During the second episode damaging floods affected the Welsh Valleys and 164mm of rain fell in 44 hours at Treherbert (Glamorgan).

1995

17–21 January   *Severe gales*
A severe gale swept the entire UK on the 17th, causing difficulties to road and rail traffic, and major delays to air and sea transport;

highest gust was 94mph at Brixham (Devon). Southern Britain had further damaging winds on the 19th with a gust to 91mph at Langdon Bay, near Dover, and again on the 21st.

25 January   *Heavy snow*
Six people died in road accidents as heavy snow affected northern England; 30cm fell in the Leeds district and up to 40cm on the Yorkshire flank of the Pennines.

30–31 January   *Severe flooding; landslips*
Cumbria and the northern Pennines were badly hit by flooding following a 22-hour downpour which deposited 225mm at Honister Pass and 167mm at The Nook, Thirlmere. Several landslips were reported, one of which was responsible for a serious railway accident at Kirkby Stephen.

17 March   *Severe gale*
There were many reports of structural damage in Wales, the Midlands and East Anglia caused by high winds. Gusts of 70–90 mph were reported inland and one of 97mph at Cardiff.

28 March   *Heavy snow*
Heavy snow blocked roads in Yorkshire and Lancashire; 35cm was reported at Holmfirth.

April to August   *Drought*
An acute shortage of rain over a relatively short period – five months – led to one of the most reprehensible 'droughts' on record. Yorkshire Water was forced to ferry water to tens of thousands of customers in tankers, but the water shortage was more a reflection of Yorkshire Water's management of its resources than it was of a real drought. The company's board subsequently resigned en masse. Over England and Wales, only 153mm of rain fell during the five-month period, just 45 per cent of the normal amount, but the fact remains that such a short dry spell should not have occasioned such a catastrophic failure of supply.

1–11 September   *Severe flooding*
Repeated heavy downpours affected northern and eastern Scotland

and north-east England, leading to widespread flooding. The floods were catastrophic in Banffshire, Morayshire and Nairnshire, and around Inverness, with thousands of acres of crops ruined, hundreds of homes inundated and road and rail arteries cut. The Aberdeen to Inverness railway line was out of action for several weeks after being washed out in several places. At Kinloss, 272mm of rain fell over the eleven-day period.

5–6 December   *Heavy snow*
Moderate snowfalls affected south-east England, with heavy falls – around 15cm – in west Kent and east Surrey. At Meopham, Kent, 20cm was measured. Three thousand motorists were stranded all night on the Kent–Surrey section of the M25.

24–31 December   *Snowstorms; freezing rain; severe cold*
This was a period of almost unprecedented cold in Scotland, and, according to government figures, some 1,400–1,500 people, mostly elderly, perished from hypothermia. The temperature fell to –27.2°C at Altnaharra, Sutherland, early on the 30th, while the previous day Glasgow airport reported an early-morning minimum of –20°C and an afternoon maximum of –12°C. There was no widespread snowfall, but severe blizzards affected Shetland on the 24th–25th (Butt of Lewis recorded a gust of 99mph and Fair Isle one of 94mph during this snowstorm), with undrifted snow 35cm deep and drifts to 9m, while Scarborough was cut off by a 35cm fall on the 25th–26th. Freezing rain fell for 12–15 hours across much of central and southern parts of both England and Wales on the 30th leading to severe dislocation of road and rail transport over a wide area. As a thaw set in on the 31st many thousands of homes experienced burst pipes.

1996

25–28 January   *Heavy snow*
Light to moderate snow was widely reported, but snow was heavy and prolonged in north-east England and south-east Scotland, blocking most high-level and some low-level roads. Snow-depths between 20 and 40cm were typical, with 50cm at Boltshope Park in the north-western corner of County Durham.

5–6 February *Snowstorm*
Heavy drifting snow brought much of north-west England and south-west Scotland to a standstill, and on the M74 hundreds of motorists were trapped for over 24 hours. Eskdalemuir, Dumfriesshire, reported 50cm of level snow, while at Boltshope Park 35cm of fresh snow took the total depth there to 70cm. It was probably the deepest snowfall in the Lancaster–Morecambe–Barrow area since 1940.

19–20 February *Heavy snow; gale; storm surge*
Heavy snow, drifting readily in the gale-force northerly wind, amounted to 10–20cm in parts of Lincolnshire and East Anglia. The northerly gale also drove a major North Sea storm surge down the east coast of England, resulting in serious coastal flooding, notably in East Yorkshire where the Spurn isthmus was breached, in Lincolnshire and Norfolk, and at Deal in east Kent.

19 May *Heavy rain; gales*
A true disaster was narrowly averted when, in atrocious conditions, the Ten Tors training competition in Devon was abandoned halfway through, and competitors were rescued by helicopter. Dartmoor streams turned into raging torrents as 50–70mm of rain fell, and a combination of high winds and low temperatures led to concerns about survival.

7 June *Thunderstorms; heavy hail; tornado*
A brief but intense heatwave broke on the evening of the 7th when thunderstorms, accompanied by giant hail, developed in a zone extending from Dorset and Wiltshire to Cambridgeshire and Norfolk. Hailstones 2–4cm across were widely reported in this zone, causing much damage to glasshouses and motor vehicles, and at Cambridge stones were said to be 5–6cm in diameter. Dozens of homes were struck by lightning, and ball lightning caused minor damage at Tewkesbury. At Wantage, 74mm of rain fell, and a tornado was seen in Basingstoke.

12 August *Thunderstorm; severe flood*
At Folkestone 112mm of rain was recorded, mostly in six hours, and large parts of the town centre were under 1–2m of water.

28–29 August   *Flooding*
Heavy rain in Norfolk caused flooding in the area between Norwich and Sheringham; at Coltishall 102mm of rain fell.

27 October   *Severe gale*
Former Hurricane 'Lili' swept across southern Britain, the wind gusting to 92mph at North Hessary Tor in Devon. Structural damage and coastal flooding were widely reported.

5–6 November   *Severe gale*
Damage to buildings, power lines and woodland affected Northern Ireland, western and central Scotland, and Cumbria. Peak gusts included 94mph at Tiree and, in Ireland, 103mph at Malin Head.

19–20 and 24–25 November, 2 and 3 December   *Heavy snow*
Three deep depression tracking north-eastwards across the middle of the UK brought snow to many parts of the country, but heavy and disruptive falls occurred on all three occasions in upland parts of north-east England and south-east Scotland. Up to 30cm fell each time, with thaws between.

27–31 December   *Heavy snow*
Heavy falls were confined to Kent and east Sussex, with 20–25cm locally.

1997

24 February   *Severe gale; storm surge*
A severe westerly gale blew across most of the UK, but winds were strongest – and damage reports most extensive – in southern counties of England where a storm surge in the English Channel also led to a good deal of coastal flooding. Lee-on-Solent logged a gust of 91mph.

10 March   *Thick fog*
Six people died in a motorway pile-up on the M42 near Bromsgrove in thick fog that morning.

9–21 May   *Thunderstorms; heavy hail; tornadoes*
This was an exceptionally thundery period, and the 17th was

probably the worst day with damage from hailstones up to 3cm across in Buckinghamshire and Bedfordshire. At Woburn (Bedfordshire) 86mm of rain fell in two hours, and a tornado caused damage in nearby Wootton.

30 June to 1 July   *Severe flood*
Hundreds of homes were flooded, farmland inundated and road and rail routes cut in Nairnshire and Morayshire, less than two years after an even more damaging flood. Elgin was cut off for a time. At Relugas (Morayshire), over 120mm fell in 48 hours, with 90mm on the 1st alone.

3–4 August   *Severe flood*
South Cornwall was badly hit by flooding following a prolonged downpour which deposited 132mm in 48 hours at Stithians, 125mm at Wendron and 100mm at Culdrose.

19 August   *Severe flood*
Floods and mudslips affected parts of north-west London, south-west Hertfordshire and neighbouring parts of Buckinghamshire. At Chorleywood, 114mm of rain fell in 100 minutes.

24 December   *Severe gale*
Severe gales swept the whole country, but the wind was especially damaging in Wales, north-west England, south-west Scotland and Northern Ireland, where structural damage and power failures were widely reported. There were six deaths. Tens of thousands of people spent a joyless Christmas in the dark and cold, and were not finally reconnected until the 27th or 28th. The highest gust speed was 112mph at Aberdaron (Caernarfon).

1998

1–8 January   *Severe gales; storm surge; tornado*
This was an exceptionally rough period over southern England and south Wales with frequent gales and a good deal of coastal flooding in areas bordering the English Channel. The most severe gale was that of the 4th when gust-speeds reached 115mph at Mumbles (near

Swansea), 105mph at Portland (Dorset) and 102mph at Davidstow Moor (Cornwall). Several tornadoes were reported, the most damaging of which struck Selsey (Sussex) just before midnight on the 6th; some 500–600 houses were damaged.

8–9 April   *Severe flooding*
Between 60 and 90mm of rain fell in 48 hours (much of it in twelve hours) in a broad zone extending from Breconshire across the south Midlands to Cambridgeshire. The highest total was 105mm near Shipston-on-Stour, Wawickshire. This was probably the most destructive flooding episode in the area since July 1968, and the rivers Avon, Leam, Cherwell and Nene all overtopped their banks. Leamington Spa, Warwick and Northampton were all very badly hit; 2,000 homes in Northampton alone were flooded. Six people died.

11–15 April   *Heavy snow*
Several rather localized but nonetheless heavy falls of snow occurred over the Easter period. On the 11th, 30cm lay at Cynwyd (Merionethshire) and on the 15th, 30cm at Moel-y-Crio and 37cm at Mold (both Flintshire). Appreciable depths were also reported from Northern Ireland, the Isle of Man, north-west England and the north-west Midlands.

20–27 October   *Severe flooding*
Repeated and prolonged bouts of rain caused widespread flooding in south Wales and parts of the West Country, and also in the floodplains of the Severn, Wye and Monnow. The Welsh Valleys were especially hard hit with hundreds of homes affected. At Treherbert in the Rhondda valley a total of 319mm of rain was recorded over an eight-day period.

26–29 December   *Severe gale; avalanche*
A destructive gale blasted across practically the entire UK, with central and southern Scotland, Northern Ireland, northern England and north Wales (roughly the same areas which suffered the previous Christmas) worst hit. Structural damage occurred widely, and thousands were left without power for two or more days. Highest gusts were 103mph at Prestwick (Ayrshire), 100mph at Capel Curig (Snowdonia) and 94mph at Machrihanish (Kintyre).

Five people died in the gale, and a further four died in an avalanche on Aonach Mor (near Ben Nevis) on the 29th.

1999

4–5 January *Severe flooding*
Serious floods in many parts of Cumbria and north Lancashire, and also along the Rivers Tyne, Wear and Tees. At Honister Pass in the Lake District 191mm of rain fell in 22 hours.

5–11 February *Heavy snow*
Road, rail and air transport in north-east Scotland was disrupted following a series of snowfalls which left 27cm of snow on the ground at Aberdeen airport, and 30–40cm elsewhere in Aberdeenshire.

5–6 March *Severe flooding*
Over 100mm of rain fell in 48 hours over the North York Moors which, allied to rapidly melting snow, produced extensive flooding, notably in Malton, Norton and Pickering.

2–5 July *Thunderstorms; local flooding*
Widespread thundery activity with locally heavy hail and a few tornado reports occurred at the beginning of July. Mid- and west Essex were hit by flooding; 122mm fell at Sible Hedingham on the 4th.

8–9 August *Severe flooding*
Thundery downpours led to severe local flooding in places. 117mm of rain fell at Windsor.

18–19 September *Severe flooding*
Prolonged heavy rain deposited 92mm in 24 hours at Newport (Monmouthshire).

4 November *Severe flooding*
Whitehaven (Cumbria) suffered much flooding when 90mm of rain fell in just over ten hours.

3 December   *Severe gale*
Hundreds a buildings were damaged by high winds, high-sided vehicles blown over on motorways and power lines brought down; highest gusts were 96mph at Capel Curig and 92mph at Aberdaron.

17–19 December   *Severe flooding*
Floods in many parts of Cornwall followed three days of heavy rain; 101mm fell at Falmouth.

24–25 December   *Severe gale; coastal flooding*
Gales swept the UK for the third consecutive Christmas holiday. A gust of 99mph was recorded at Plymouth. High tides and a storm surge led to flooding along English Channel and Bristol Channel coasts.

2000

3 January   *Severe gale*
Damaging winds blasted northern Scotland with gusts to 106mph at Lerwick and Stornoway.

27–29 May   *Flooding*
The late May bank holiday weekend was a washout, especially in East Anglia and south-east England; at Clacton (Essex) 100mm of rain fell in 72 hours.

4 July   *Flooding*
Heavy thunderstorms caused flooding in Kent, Surrey, Sussex and Hampshire. At Midhurst (Sussex) 96mm of rain fell, mostly in less than six hours.

15–16 September   *Flooding*
Almost 100 homes were flooded in Portsmouth after a night of heavy rain, with 84mm falling at Walderton (Sussex). The flood was exacerbated by the failure of a pumping station.

11–12 October   *Severe flood*
In a prelude to the forthcoming winter of record-breaking floods, disastrous flooding affected towns and villages in Sussex and Kent on the rivers Ouse, Cuckmere, Rother, Teise, Beult and Medway. Up

to 3m of floodwaters tore through the streets of Lewes which was cut off for almost two days. There were serious floods, too, in parts of Brighton and Eastbourne. At Plumpton, just north-west of Lewes, 174mm of rain was recorded in 48 hours, of which almost 120mm fell in twelve hours.

30 October *Severe gale; storm surge; tornado*
A ferocious gale lashed southern England and south Wales, bringing down hundreds of trees, hundreds of kilometres of power and telephone lines, and causing a good deal of damage to homes and other buildings. With heavy rain an additional hazard, road and rail transport was thrown into chaos. Gusts included 97mph at Mumbles (near Swansea), with 85–95mph at many other sites in south Wales and the West Country. A tornado caused structural damage in Bognor. Eight people died during the bad weather on this day.

October 2000 to March 2001 *Widespread and prolonged flooding*
The autumn and winter of 2000–1 established new England and Wales rainfall records for every period from three months upwards, and the autumn quarter was a full 10 per cent wetter than the previous record-holder – 1852. In East Sussex more than three times the normal amount of rain fell: in other words, ten months' worth of rain fell there between September and November. Widespread heavy rain amounting to 60–90mm over most central and southern parts of both England and Wales between the 29th and 31st of October was followed by a similar fall during the opening days of November which affected all parts of England and Wales. The inevitable floods which followed have few, if any, precedents in British climatological history; on the Severn, the Thames, the Great Ouse and the Trent they were legitimately described as the worst since 1947, and at York water levels rose higher than on any other occasion in the previous 400 years. Damaging floods also returned to Sussex and Kent. The peak of the floods passed during the second week of November, but renewed heavy rains which recurred at frequent intervals until March meant that the floodwaters never completely subsided, and many places, notably in Sussex, endured five or six separate flooding episodes during this appalling season.

13–14 December   *Severe gale*
Much damage and disruption occurred over England and Wales during this gale. Highest gusts included 93mph at Mumbles (Glamorgan) and 90mph at Cranwell (Lincolnshire).

27–28 December   *Heavy snow*
Moderate falls of snow were widely reported, but heavy snow affected Northern Ireland and western Scotland. In both Glasgow and Belfast snow lay 20cm deep, and falls of 25–35cm were reported from parts of Lanarkshire, Ayrshire and County Antrim.

2001–8

2001

4–7 February   *Heavy snow*
Heavy snow fell across much of Scotland, Northern Ireland and northern England, blocking roads and railways and closing airports. At low levels the snow was soft and clinging, and brought down power lines. A train was trapped in drifts on the line from Inverness to Wick, and over 11,000 homes were without electricity. Snow depths (undrifted) of 25–40cm were reported widely, and at Aboyne (Aberdeenshire) it was 60cm deep.

26 February to 3 March   *Heavy snow*
Deep snow was less widespread than during the event three weeks earlier; nevertheless 50–60cm fell in the Lanark–Carluke district, 33cm was measured at Boltshope Park (County Durham), 27cm at Lerwick (Shetland) and 15cm at Norwich.

3–7 July   *Thunderstorms; flash flood*
Severe thunderstorms accompanied by torrential downpours were widely reported during this period. A flash flood swept through a caravan site near Bala on the 4th, and although no local rainfall record is available, 105mm fell at Betws-y-Coed, some 25km to the north.

17–19 July   *Flooding*
Prolonged heavy rain led to widespread, though short-lived, flooding in the east Midlands and Norfolk. At Keyworth (near Nottingham) 96mm fell, mostly within fifteen hours, while at Upper Sheringham (Norfolk) 135mm fell in 72 hours, of which 110mm came down in fifteen hours.

21 October   *Flooding*
Flooding in Cambridge was extensive, and there was a good deal of flooding, too, along the Cam, Great Ouse and Roding. Some 90mm of rain fell in central Cambridge.

27–31 December   *Heavy snow*
During this wintry period, heavy snow was confined to northern Scotland where 25–35cm fell widely in east Sutherland, Easter Ross and inland of Inverness.

2002

28 January   *Severe gale*
Central and southern Scotland and Northern Ireland were worst hit by this gale. Damage occurred widely and seven people died. A gust of 108mph was recorded on the Forth road bridge.

1–2 February   *Severe gales; storm surge; flooding*
Structural damage, power failures and disruption of transport were compounded by exceptional storm surges around western and southern coasts which caused extensive coastal flooding. River flooding also occurred widely, notably in south Wales where Ebbw Vale collected 108mm on the 1st, and 144mm over the two days. Highest gust was 91mph at Capel Curig (Snowdonia).

22 February   *Severe gale*
A notable gale was accentuated by the lee-wave effect in northern England, causing widespread damage and driving problems in Yorkshire. A large part of the roof of York railway station was ripped off. Gusts approaching 90mph were reported in the Vale of York.

29 July to 9 August   *Severe thunderstorms; flooding*
Thunderstorms, often severe and prolonged, affected many parts of the country daily through this period and associated downpours caused many serious, though generally short-lived flooding episodes. On the 30th, 113mm fell at Penistone (South Yorks), on the 1st, Fylingdales and Leeming (both North Yorkshire) reported 115mm and 91mm respectively, and on the 7th, 71mm fell at Hampstead (London), most of it in 75 minutes during the evening rush-hour. Large parts of the London Underground were suspended for the rest of the evening.

9 September   *Severe flood*
Damaging floods affected large parts of Dorset and the Isle of Wight. Swanage recorded 122mm of rain.

27 October   *Severe gale*
High winds battered central and southern parts of England and Wales, and the most serious damage was confined to a relatively narrow zone extending across south Wales, the south Midlands and East Anglia. Power lines were damaged, and several thousand people were without electricity for 24 to 48 hours, and some for up to a week. Highest gust was 97mph at Mumbles (near Swansea), but gusts of 70–80mph were widely recorded inland too.

13–14 November   *Severe flood*
Floods occurred widely and repeatedly during this very wet month, and south Cornwall was very badly hit on these two days following a fall of 84mm at Penzance. St Ives suffered its worst flood for several decades. Flooding returned to the same area on 27–28 November.

27 December to 2 January 2003   *Severe flood*
All southern counties of England were affected by flooding which was particularly disruptive in Cornwall, south Devon, along the Thames, and in Surrey and Kent. At Cardinham (Cornwall) 225mm fell in two weeks starting 21 December, and Kenley (Surrey) collected 99mm in six days starting 28 December. The Thames at Oxford was at its highest since 1947.

2003

30–31 January Heavy snow
Only moderate snow depths were reported, but the coincidence of heavily falling snow with the afternoon rush-hour in the London area led to a complete gridlock of the capital's roads, while thousands of drivers on the M11 spent a night stranded on the motorway. Some were stuck for almost eighteen hours. Depths were typically 5–12cm, although 20cm was reported from Lakenheath (Suffolk). Some 15–20cm of snow also fell in north-east Scotland, and further falls of 25–30cm followed in the Scottish Highlands between the 3rd and 5th February.

2 to 13 August *Heatwave; thunderstorms; heavy hail; flooding*
The temperature closely approached or exceeded 32°C somewhere in England on nine consecutive days, passing 35°C on the 6th, 9th and 10th. The highest reading of 38.1°C at Kew Gardens constituted a new all-time UK-wide record. Mortality during the heatwave exceeded the norm by slightly more than 2,000, with the vast majority of the victims being elderly people suffering from dehydration and heatstroke; however, the Government's official assessment places the death toll at just over 900. There were also severe local thunderstorms which, for instance, dropped 93mm in ten hours on Colonsay in the Inner Hebrides on the 5th and 49mm mostly in twelve minutes at Carlton-in-Cleveland, just south of Middlesbrough. Heavy hail and short-lived but dramatic flooding affected large parts of the Teesside district.

21–26 November *Flooding*
A prolonged downpour lasting over 60 hours dropped 80–100mm of rain across a large part of London and the Home Counties. Charlwood (Surrey) collected 125mm in six days.

2004

27–28 January *Heavy snow; thunder*
A marked squall-line brought a brief period of heavy snow accompanied by thunder and very gusty winds moved steadily southwards

across the entire UK, causing considerable problems for road and air traffic. Birmingham was grid-locked for several hours. Snow depths were mainly small, but 15cm fell at Coningsby (Lincolnshire). In upland regions, 30cm lay at Fylingdales and 28cm at Glenlivet.

30 January to 5 February   *Severe flood*
Serious flooding affected the various river systems draining Snowdonia (there were lesser floods in Cumbria and south-west Scotland) following prolonged orographic rainfall. The Conway catchment, especially around Llanrwst, was worst hit. At Capel Curig, 169mm was recorded in 24 hours, 274mm in 72 hours and 418mm in a week.

20 March   *Severe gale*
A damaging gale swept England and Wales; highest gust was 100mph at Aberdaron (Caernarfon). Four deaths were reported.

7–8 July   *Thunderstorms; flooding*
RAF Wittering, near Peterborough, collected 109mm in just over 24 hours in two separate falls. There was a good deal of flooding in the district, including the airfield where lightning damage also occurred.

9–18 August   *Widespread thunderstorms; mudslides, flash flood*
The 'Boscastle disaster' struck the north Cornish resort on the afternoon of the 16th when a flash flood tore through the narrow valley of the River Valency in response to a four-hour deluge which dropped 200mm of rain at Otterham, on the moor above Boscastle. It is estimated that 250mm may have fallen at the centre of the downpour. More than 70 properties were damaged, several irreparably, 50 vehicles were washed into the harbour, and over 50 people had to be rescued, many of them by helicopter. It was a miracle that no one died. There were a large number of other major downpours triggering floods and landslips during this period, notably on the 18th at Lochearnhead in the Southern Highlands where dozens of motorists were trapped by two mudslides. The week before, nearby Aberfeldy collected 143mm of rain in four days, while West Wittering (near Selsey, in Sussex) measured 91mm in a storm on the 10th.

27 October *Storm surge*

A south-easterly gale, not particularly noteworthy for its intensity (highest gust 67mph at Brixham) was associated with a massive southerly swell which also coincided with spring tides to deliver the worst coastal flooding in south Cornwall since March 1962. Dozens of small vessels at anchor in supposedly sheltered harbours were sunk.

2005

6–12 January *Severe gales; severe floods*

Two very deep depressions were responsible for this exceptionally rough period. The first of these disturbances crossed Scotland on the 8th with extensive damage reported from all parts of England, Wales and southern Scotland; the highest gust was 103mph at St Bee's Head (Cumbria). A prolonged orographically enhanced rainfall affected mid- and north Wales, Cumbria and the northern Pennines, and the Southern Uplands, and the resulting flood in Carlisle was one of the worst in the town's history with at least 2,000 homes and business premises inundated. Capel Curig (Snowdonia) recorded 144mm in 24 hours and 184mm in 48 hours, Keswick (Cumbria) had 118mm in 48 hours, and Shap (Cumbria) 227mm in 72 hours. The second depression tracked past north-west Scotland, and a maximum gust of 106mph was recorded at Barra (Western Isles). This south-westerly gale was responsible for a storm surge which caused flooding in the Hebrides and along the west coast of the Scottish mainland. The death toll for the week approached twenty.

24 February to 4 March *Heavy snow*

A snowstorm on the 24th–25th which lasted almost 24 hours affected chiefly upland districts of the north Midlands, northern England and southern Scotland, but here its effects were very severe, blocking roads, bringing down power lines and forcing hundreds of school closures. By the morning of the 25th, snow lay 52cm deep at Boltshope Park (County Durham) with 6m drifts; the snow cover here lasted until 14 March. There were frequent further falls, mostly light to moderate, during the remainder of February, but heavy snow fell in Kent and East Sussex on the 27th and again on 1–2 March, depositing 30cm over the North Downs between Ashford and Dover, and some 20–25cm in and around Canterbury.

19 June   *Thunderstorms; heavy hail; flash flood*
Thunderstorms broke out widely during the afternoon and evening in Wales, the Midlands and northern England, some accompanied by damaging hail. A torrential downpour over the North York Moors led to a serious flash flood in Ryedale, with many homes, roads, bridges and campsites suffering damage. At Hawnby, 69mm of rain was recorded, mostly in 30 minutes, but it is estimated that over 100mm fell at the focus of the storm. Hailstones up to 4 or 5cm in diameter were noted both in North Yorkshire and in the Black Country.

28 July   *Thunderstorms; tornado*
A destructive tornado tore through south-east Birmingham during the afternoon, affecting very nearly the same part of the city as was hit in June 1931. The tornado touched down in Kings Heath, travelled across Moseley, Balsall Heath and Small Heath, and finally lifted off the ground in Erdington. The track was 500m wide during the most intense phase in Balsall Heath. More than 20 people were injured, three seriously, and scores of houses were badly damaged. Repair costs were estimated at £40–50 million. Another significant tornado was reported from the Peterborough district, and thunderstorms occurred widely.

10 September   *Severe thunderstorm*
Storms broke out widely across southern England; 87mm fell in two hours at Chieveley (Berkshire).

11–12 October   *Severe flood*
A good deal of flooding occurred in south-west Scotland, north-west England and Wales after a prolonged downpour; Milford Haven (Pembrokeshire) collected 120mm in 24 hours and 140mm in 48 hours; other 48-hour totals were 113mm at both Eskdalemuir and Carlisle.

25–28 November   *Heavy snow*
Snow fell heavily across Scotland early on the 25th, with 20cm at Glenlivet and an estimated 30–40cm in the hill-country of north Aberdeenshire. Later in the day there were falls of 15–20cm in mid- and west Glamorgan, and in Cornwall, where 900–1,000 motorists

were stranded on the A30 on Bodmin Moor overnight. Further moderate snow disrupted road and air transport on the 28th, closing Manchester airport for a time, and blocking roads in Gloucestershire.

26–30 December   *Heavy snow*
Snow fell widely, but heavy falls were confined to east Kent on the 26th–28th, totalling 30cm over the North Downs. Drifting snow in Lincolnshire and Yorkshire caused serious problems for motorists on the 30th ahead of a general thaw.

November 2004 to August 2006   *Drought*
An extended period of low rainfall affected the southern halves of England and Wales over this 22-month period, resulting in widespread restrictions on the use of water, although there remains considerable doubt as to whether all of the restrictions were entirely necessary. The deficit over the entire period exceeded 25 per cent only in a very narrow zone extending from north Hampshire and south-west Berkshire to west Kent.

2006

28 February to 13 March   *Heavy snow*
A northerly outbreak in the first week of March delivered 15–25cm of snow to much of northern and eastern Scotland and north-east England; Glenlivet had 30cm. After a thaw, heavy snow returned to north Wales, north-west England and western Scotland overnight 12th–13th, with up to 40cm in the higher parts of the Glasgow area.

26 June   *Flooding*
Penzance recorded 82mm of rain, much of it in two hours, resulting in serious but short-lived flooding in south Cornwall.

1–31 July   *Heatwave; thunderstorms*
Over a large part of England this was the hottest calendar month (though not the hottest 30-day period, q.v. 1976 and 1995) on record, the temperature exceeding 32°C on eight days after midmonth. The mortality rate approached 1,000 above normal. Severe

thunderstorms, often accompanied by heavy hail, occurred frequently; associated downpours included 89mm at Chippenham (Wiltshire) on the 5th and 75mm at Charlbury (Oxfordshire) on the 22nd.

13 August   *Thunderstorms; flooding*
Serious flooding affected west Surrey and neighbouring parts of Berkshire and Hampshire following a series of thunderstorms; at Virginia Water 118mm of rain was measured.

25 September   *Thunderstorms; flooding*
A serious flood damaged property in Great Yarmouth; 117mm of rain fell at nearby Gorleston.

25 October   *Severe gale; flooding*
A severe north-easterly gale raged across northern Scotland, with gusts to 89mph at Wick and 85mph at Lerwick. A trawler sank in high seas with the loss of its crew of five. Prolonged heavy rain caused widespread flooding in Sutherland, Caithness and Orkney; 117mm of rain fell at Rackwick (Orkney) and 98mm at Altnaharra (Sutherland).

7 December   *Tornado*
A tornado caused a good deal of structural damage in Kensal Rise, north-west London.

19–23 December   *Fog*
Widespread fog, which persisted for four days with only occasional intermission, affected road and air travel over much of England and Wales. The pre-Christmas rush at many airports resulted in serious delays for thousands of travellers (see Chapter 5).

2007

6–10 January   *Severe flood*
Heavy orographic rains affected Wales, north-west England and western Scotland, feeding floods on the Tay and Clyde, on the Eden and on the Conway. At Capel Curig, 245mm of rain was recorded over the five-day period.

18 January *Severe gale*
A fierce westerly gale swept the whole of England and Wales, causing mayhem on the roads, uprooting thousands of trees and causing a considerable amount of property damage. In all, nineteen people died in the UK as a consequence of the bad weather. Atlhough the peak gusts of 84mph at Crosby and Squires Gate (both Lancashire) were not abnormally high, damaging gusts of 70–80mph were recorded widely at inland sites throughout England (see Chapter 7).

8–10 February *Heavy snow*
Widespread snow disrupted traffic and resulted in the closure of thousands of schools. Undrifted depths included 16cm at Chesham (Buckinghamshire) on the 8th, 27cm at Boltshope Park (Durham) on the 9th and unconfirmed reports of 30–40cm in mid-Wales also on the 9th.

May to July *Repeated and widespread severe flooding*
Exceptional amounts of rain fell in each of these three months, and there were seven episodes of severe flooding. On 13–14 May Shrewsbury and neighbouring parts of Shropshire were badly affected; 79mm of rain fell at Preston Montford. Over the late spring bank holiday weekend, 26–28 May, flooding was reported from the Isle of Wight across the Home Counties to East Anglia, with Luton suffering most, as a result of 99mm of rain falling in 48 hours. Between 13th and 15th June, 112mm fell in 48 hours at Edgbaston in Birmingham, 123mm at Wilsden, near Bradford, in 60 hours and 123mm at Helen's Bay (Co. Down) in 84 hours. Serious flooding occurred in Birmingham, South and West Yorkshire and in Counties Antrim and Down. On 19 June a violent thunderstorm led to a destructive flash flood at Hampton Loade, just south of Bridgnorth, in Shropshire. There was further prolonged heavy rain on 24–25 June over the same areas which suffered ten days earlier, and this time catastrophic flooding hit Hull, Sheffield, Rotherham and Doncaster; 112mm of rain fell in 48 hours at Fylingdales (North Yorkshire). The focus of attention moved to Lancashire during the first week of July, with Stonyhurst, north of Blackburn, collecting 115mm of rain in six days. But the biggest downpour of the entire summer struck the south-west Midlands and upper Thames Valley during 19 and 20 July; 48-hour totals included 157mm at Pershore and 128mm at

Brize Norton; and more than 100mm of rain fell over an area of about 2,500 sq. km. The floods along the River Severn from north of Worcester to beyond Gloucester were considered to be the worst since 1947, and the worst during the summer half-year since May 1886. The Avon, Teme, Wye, Bristol Avon, Thames, Cherwell and Kennett also suffered serious flooding (see Chapter 6).

8–9 November   *Severe gale; storm surge*
A north-westerly gale in northern Scotland produced a peak gust of 93mph at Fair Isle, and it triggered a significant storm surge along North Sea coasts. However, the peak of the surge was not in phase with the high tide, and – notwithstanding dire warnings from the Environment Agency – coastal flooding was very limited.

2008

18–21 January   *Severe flood*
Prolonged orographic rain affected the hills of Wales, northern England and south-west Scotland, and dropped 213mm in 84 hours at Capel Curig (Snowdonia). Serious flooding affected the Conway and Eden valleys, and also the River Aire, with parts of Leeds badly hit. A few days later further rains over western Scotland triggered floods on the Tay and Tummel.

29 May   *Flooding*
Heavy downpours in Devon, Dorset and Somerset led to serious but short-lived flooding, notably in the Torbay, Ilminster and Bruton districts. Some 70–80mm of rain fell in four hours.

Index